153
Advances in Polymer Science

Editorial Board:
A. Abe · A.-C. Albertsson · H.-J. Cantow · K. Dušek
S. Edwards · H. Höcker · J. F. Joanny · H.-H. Kausch
T. Kobayashi · K.-S. Lee · J. E. McGrath
L. Monnerie · S. I. Stupp · U. W. Suter
E. L. Thomas · G. Wegner · R. J. Young

Springer
Berlin
Heidelberg
New York
Barcelona
Hong Kong
London
Milan
Paris
Singapore
Tokyo

Biopolymers · PVA Hydrogels Anionic Polymerisation Nanocomposites

With contributions by
J.Y. Chang, D.Y. Godovsky, M.J. Han, C.M. Hassan,
J. Kim, B. Lee, Y. Lee, N.A. Peppas,
R.P. Quirk, T. Yoo

This series presents critical reviews of the present and future trends in polymer and biopolymer science including chemistry, physical chemistry, physics and materials science. It is addressed to all scientists at universities and in industry who wish to keep abreast of advances in the topics covered.

As a rule, contributions are specially commissioned. The editors and publishers will, however, always be pleased to receive suggestions and supplementary information. Papers are accepted for „Advances in Polymer Science" in English.

In references Advances in Polymer Science is abbreviated Adv. Polym. Sci. and is cited as a journal.

Springer APS home page: http://link.springer.de/series/aps/ or
http://link.springer-ny.com/series/aps
Springer-Verlag home page: http://www.springer.de

ISSN 0065-3195
ISBN 3-540-67313-X
Springer-Verlag Berlin Heidelberg New York

Library of Congress Catalog Card Number 61642

This work is subject to copyright. All rights are reserved, whether the whole or part of the material is concerned, specifically the rights of translation, reprinting, re-use of illustrations, recitation, broadcasting, reproduction on microfilms or in other ways, and storage in data banks. Duplication of this publication or parts thereof is only permitted under the provisions of the German Copyright Law of September 9, 1965, in its current version, and permission for use must always be obtained from Springer-Verlag. Violations are liable for prosecution under the German Copyright Law.

Springer-Verlag Berlin Heidelberg New York
a member of BertelsmannSpringer Science+Business Media GmbH

© Springer-Verlag Berlin Heidelberg 2000
Printed in Germany

The use of registered names, trademarks, etc. in this publication does not imply, even in the absence of a specific statement, that such names are exempt from the relevant protective laws and regulations and therefore free for general use.

Typesetting: Data conversion by MEDIO, Berlin
Cover: MEDIO, Berlin
Printed on acid-free paper   SPIN: 10706226   02/3020hu - 5 4 3 2 1 0

Editorial Board

Prof. Akihiro Abe
Department of Industrial Chemistry
Tokyo Institute of Polytechnics
1583 Iiyama, Atsugi-shi 243-02, Japan
E-mail: aabe@chem.t-kougei.ac.jp

Prof. Ann-Christine Albertsson
Department of Polymer Technology
The Royal Institute of Technolgy
S-10044 Stockholm, Sweden
E-mail: aila@polymer.kth.se

Prof. Hans-Joachim Cantow
Freiburger Materialforschungszentrum
Stefan Meier-Str. 21
D-79104 Freiburg i. Br., FRG
E-mail: cantow@fmf.uni-freiburg.de

Prof. Karel Dušek
Institute of Macromolecular Chemistry, Czech
Academy of Sciences of the Czech Republic
Heyrovský Sq. 2
16206 Prague 6, Czech Republic
E-mail: office@imc.cas.cz

Prof. Sam Edwards
Department of Physics
Cavendish Laboratory
University of Cambridge
Madingley Road
Cambridge CB3 OHE, UK
E-mail: sfe11@phy.cam.ac.uk

Prof. Hartwig Höcker
Lehrstuhl für Textilchemie
und Makromolekulare Chemie
RWTH Aachen
Veltmanplatz 8
D-52062 Aachen, FRG
E-mail: hoecker@dwi.rwth-aachen.de

Prof. Jean-François Joanny
Institute Charles Sadron
6, rue Boussingault
F-67083 Strasbourg Cedex, France
E-mail: joanny@europe.u-strasbg.fr

Prof. Hans-Henning Kausch
Laboratoire de Polymères
École Polytechnique Fédérale
de Lausanne, MX-D Ecublens
CH-1015 Lausanne, Switzerland
E-mail: hans-henning.kausch@lp.dmx.epfl.ch

Prof. Takashi Kobayashi
Institute for Chemical Research
Kyoto University
Uji, Kyoto 611, Japan
E-mail: kobayashi@eels.kuicr.kyoto-u.ac.jp

Prof. Kwang-Sup Lee
Department of Macromolecular Science
Hannam University
Teajon 300-791, Korea
E-mail: kslee@eve.hannam.ac.kr

Prof. James E. McGrath
Polymer Materials and Interfaces Laboratories
Virginia Polytechnic and State University
2111 Hahn Hall
Blacksbourg
Virginia 24061-0344, USA
E-mail: jmcgrath@chemserver.chem.vt.edu

Prof. Lucien Monnerie
École Supérieure de Physique et de Chimie
Industrielles
Laboratoire de Physico-Chimie
Structurale et Macromoléculaire
10, rue Vauquelin
75231 Paris Cedex 05, France
E-mail: lucien.monnerie@espci.fr

Prof. Samuel I. Stupp
Department of Measurement Materials Science
and Engineering
Northwestern University
2225 North Campus Drive
Evanston, IL 60208-3113, USA
E-mail: s-stupp@nwu.edu

Prof. Ulrich W. Suter
Department of Materials
Institute of Polymers
ETZ, CNB E92
CH-8092 Zürich, Switzerland
E-mail: suter@ifp.mat.ethz.ch

Prof. Edwin L. Thomas
Room 13-5094
Materials Science and Engineering
Massachusetts Institute of Technology
Cambridge, MA 02139, USA
E-mail. thomas@uzi.mit.edu

Prof. Gerhard Wegner
Max-Planck-Institut für Polymerforschung
Ackermannweg 10
Postfach 3148
D-55128 Mainz, FRG
E-mail: wegner@mpip-mainz.mpg.de

Prof. Robert J. Young
Manchester Materials Science Centre
University of Manchester and UMIST
Grosvenor Street
Manchester M1 7HS, UK
E-mail: robert.young@umist.ac.uk

Advances in Polymer Science
Now Also Available Electronically

For all customers with a standing order for Advances in Polymer Science we offer the electronic form via LINK free of charge. Please contact your librarian who can receive a password for free access to the full articles. By registration at:

http://link.springer.de/series/aps/reg_form.htm

If you do not have a standing order you can nevertheless browse through the table of contents of the volumes and the abstracts of each article at:

http://link.springer.de/series/aps/

There you will find also information about the

– Editorial Bord
– Aims and Scope
– Instructions for Authors

Contents

Polynucleotide Analogues
M.J. Han, J.Y. Chang   1

Structure and Applications of Poly(vinyl alcohol) Hydrogels
Produced by Conventional Crosslinking or by Freezing/Thawing Methods
C.M. Hassan, N.A. Peppas 37

Applications of 1,1-Diphenylethylene Chemistry in Anionic Synthesis
of Polymers with Controlled Structures
R.P. Quirk, T. Yoo, Y. Lee, J. Kim, B. Lee 67

Device Applications of Polymer-Nanocomposites
D.Y. Godovsky 163

Author Index Volumes 101–153 207

Subject Index 219

Polynucleotide Analogues

Man Jung Han[1], Ji Young Chang[2]

[1] Department of Molecular Science and Technology, Ajou University, Suwon 442-749, Korea, e-mail: mjhan@madang.ajou.ac.kr
[2] School of Materials Science and Engineering, Hyperstructured Organic Materials Research Center, Seoul National University, Seoul 151-742, Korea

Nucleic acids (DNAs and RNAs) are biopolymers, called polynucleotides as polymers of nucleotides. They have purine or pyrimidine bases and sugar rings connected by phosphodiester linkages. Polynucleotide analogues (PNAs) with modified polymer backbones and/or modified nucleic acid bases have been studied as model polymers for natural nucleic acids. In spite of their structural simplicity, PNAs have shown the property of base-stacking along the synthetic polymer backbones and even interacted with natural nucleic acids through base-pairing. Since biological activities of natural nucleic acids are indebted greatly to base-base interactions, PNAs are expected to have certain types of biological activities. In this article the preparation methods, properties, and potential applications of PNAs are discussed.

Keywords. Polynucleotide analogue, Depurination, Depyrimidination, Polypeptide nucleic acid, Catalytic activity

1	Introduction	2
2	Synthesis and Physicochemical Properties	3
2.1	Vinyl Polymers Containing Nucleic Acid Bases	3
2.2	Functional Polymers Coupled with Nucleic Acid Base Derivatives	3
2.3	Alternating Copolymers of Nucleoside Derivatives and Alkenyl Anhydride Monomers	5
2.3.1	Physicochemical Properties	8
2.3.2	Base Stacking.	10
2.3.3	Base Pairing	13
2.4	Polypeptides with Nucleic Acid Base Derivatives	16
3	Alteration of Polynucleotide Analogues	20
3.1	Dimerization of Nucleobases by UV Irradiation	20
3.2	Depurination and Depyrimidination	22
4	Catalytic Activity for Phosphodiester Hydrolysis and DNA Cleavage.	24

5	Application of Polynucleotide Analogues	30
5.1	Liquid Chromatography	30
5.2	Template Polymerization	30
5.3	Photoresist	31
6	Future Perspectives	33
	References	33

1
Introduction

There has always been a great interest in the relationship of structure to function, especially in biologically active compounds. This is certainly no less true of polymeric materials, where the influence of conformation is a major concern. Nucleic acids (DNAs and RNAs) which have responsibilities for genetic replication, transcription, and translation in living systems, are the most important biopolymers in existence. Since DNA was recognized as a double helix with two antiparallel nucleic acid chains by means of X-ray crystallography in 1953 [1], a tremendous amount of work has been carried out on the structure and the chemistry of nucleic acids. In addition to research on nucleic acids themselves, their structural information has led chemists to synthesize the polymer analogues in order to understand better the effects of structure on function.

Polynucleotide analogues (PNAs), as model polymers for natural nucleic acids, were expected to have some of the important properties of the natural polymers such as intramolecular base stacking and intermolecular base pairing. There was also strong hope that these synthetic polymers would interact with natural nucleic acids to show certain types of biological activities and could be used for chemotherapy. The application of natural nucleic acids as chemotherapeutic agents was usually deteriorated by hydrolysis of phosphate linkages. If the phosphate groups in natural nucleic acids were substituted by stable linkers insusceptible to enzymatic degradation, the analogues would have potential uses as the antisense, antitemplate, and antigen compounds in molecular biology. Many of the reported analogues showed properties of base stacking or base pairing and some of them also formed double- or triple-stranded complexes with natural nucleic acids [2]. To resemble natural nucleic acids more closely, recent research focuses on achieving structural similarity of the polymer backbones. In this article we review general synthesis of PNAs and emphasize recent progress on the investigation of their properties and application.

2
Synthesis and Physicochemical Properties

Generally PNAs have modified polymer backbones (phosphate groups and sugar moieties) and/or modified nucleic acid bases. These analogues can be classified by their preparation methods:
1. Radical homopolymerizations of vinyl monomers having nucleic acid bases [2, 3]
2. Grafting of nucleic acid base derivatives onto functional polymer backbones [4–6]
3. Radical alternating copolymerizations of nucleoside derivatives with alkenyl anhydrides [7–14]
4. Stepwise reactions of bifunctional compounds having nucleic acid bases [15, 16].

The first three methods generally yield the homopolymers of high molecular weights, while the fourth method has the advantage of controlling the sequences of nucleic acid bases.

2.1
Vinyl Polymers Containing Nucleic Acid Bases

Vinyl, acryl, methacryl groups, or their derivatives were introduced to the 1-position of pyrimidine (cytosine, uracil, and thymine) or 9-position of purine (adenine, guanine, and hypoxanthine) to yield the vinyl monomers, which were polymerized with the aid of radical initiators in organic solvents (Scheme 1).

Most of these radical polymerizations proceeded very rapidly and the yields of the polymerization were very high even though the vinyl monomers contained pyrimidine or purine bases. As these PNAs did not contain sugar rings or hydrophilic groups, most of them were insoluble in water. Their properties were therefore investigated mainly in organic solvents. Synthesis of the monomer and the polymerizations were well summarized in the reviews [2, 3].

2.2
Functional Polymers Coupled with Nucleic Acid Base Derivatives

PNAs were synthesized by the coupling reactions of activated nucleic acid base derivatives with functional polymers, such as polyethylenimine [17–21] or poly-L-lysine [22, 23] (Scheme 2). Hydrophilic polytrimethylenimine [24, 25], and poly(vinyl alcohol) [26, 27] were also used as functional polymers. These grafting methods have the advantage of obtaining high molecular weight polymers but also a problem with incomplete coupling. Some of these polymers were soluble in water depending on the structures of the polymer

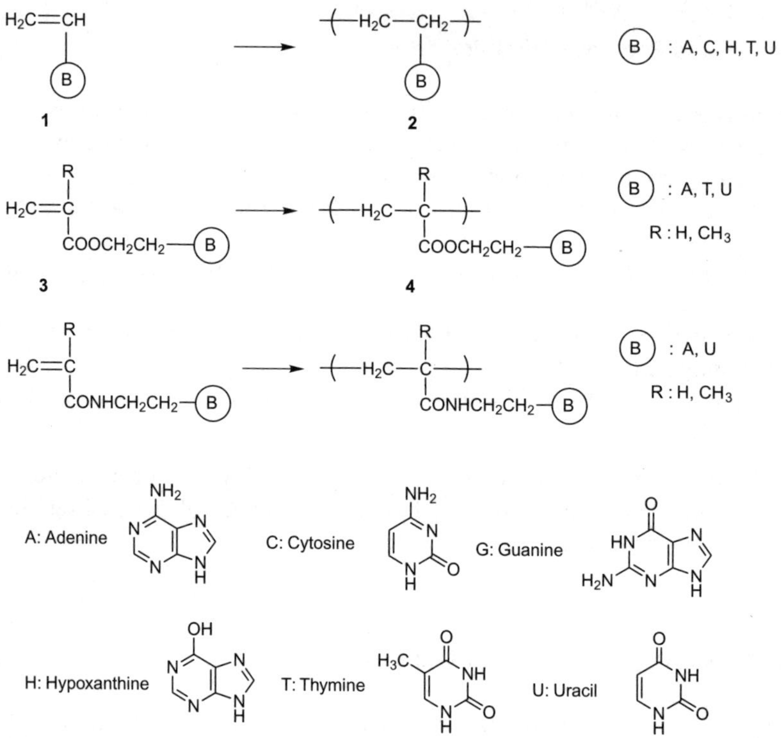

Scheme 1

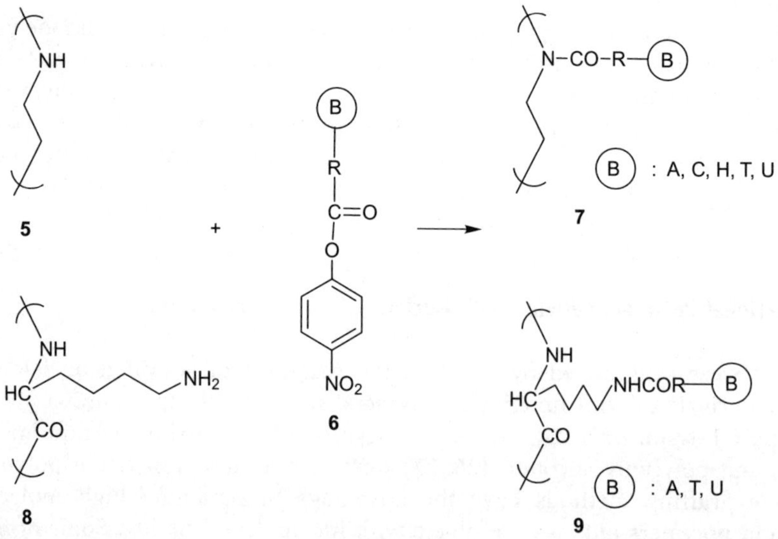

Scheme 2

backbones and of the spacers between nucleic acid bases and the polymer backbones.

Nucleic acids have a chiral center adjacent to the nucleic acid base, which is believed to provide the polymers with stereoregular structures. To confer this feature to the analogues, optically pure α-nucleic acid base substituted propanoic acids were prepared by the conventional optical resolution method. They were coupled with the functional polymers *via* amide or ester bonds to yield the optically active polymers [4–7]. Functionalization and coupling reactions of nucleobases with the polymers were well summarized in the reviews [2, 3].

2.3
Alternating Copolymers of Nucleoside Derivatives and Alkenyl Anhydride Monomers

As most of the PNAs synthesized above did not contain sugar moieties or hydrophilic groups, they were insoluble in water. Additionally, the alternating sequences between nucleoside and phosphate, observed in the natural nucleic acids, were rarely realized in the previous synthetic methods, resulting in structures quite different from those of the natural ones. One of the recent studies on PNAs intended to synthesize those resembling more closely the natural ones and investigate their properties in aqueous solutions (biological conditions) [8–14].

In order to synthesize PNAs containing ribose rings on their backbones, double bonds were introduced directly either inside of or as side groups on the ribose rings of nucleosides as follows: oxidation of deoxyribonucleosides (**10**) by platinum oxide and oxygen resulted in 5'-carboxylic acid (**11**) of the nucleosides, which gave dihydrofuran derivatives (**12**) by decarboxy dehydration with the aid of *N,N*-dimethyl formamide dineopentyl acetal [8, 10, 11]. Starting from ribonucleoside, 5'-hydroxyl group was transformed to 5'-iodo group with the aid of methyl triphenoxyphosphonium iodide. Elimination of hydrogen iodide by a strong base resulted in monomer **15** (Scheme 3) [14].

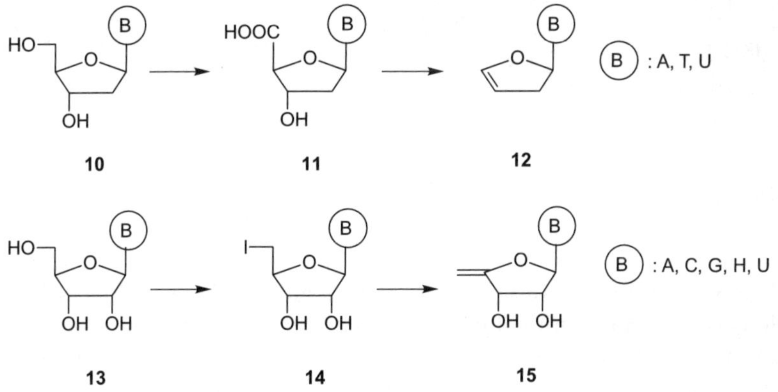

Scheme 3

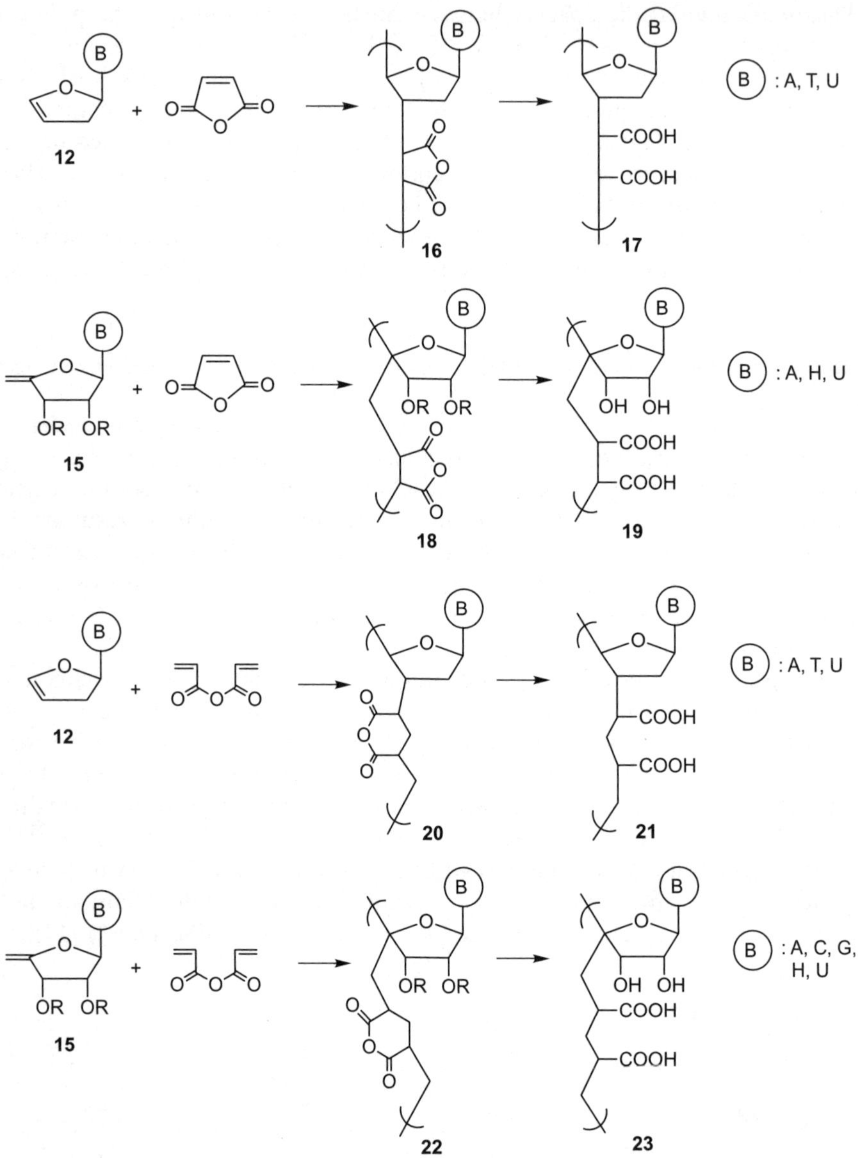

Scheme 4

Copolymerization of **12** or **15** with maleic anhydride or with acrylic anhydride gave four alternating anhydride copolymers, which were hydrolyzed to result in four PNAs (Scheme 4) [8, 10, 12, 14]. In order to facilitate the copolymerization with anhydride monomers, the primary amino groups on the bases (adenine, guanine, or cytosine) of monomers **12** and **15**, and hydroxy groups on the furanose rings of monomer **15** were blocked either by acetyl or ethoxycarbonyl groups, which could be deprotected by hydrolysis without deterioration of *N*-glycosidic bonds after the polymerization.

The radical copolymerization of vinyl ethers with maleic anhydride is known to yield alternating copolymers by forming charge-transfer complexes of the monomer pairs during the copolymerization [28–30]. As the electron-donating character of the vinyl ether groups in **12** and **15** was negligibly influenced by the base groups substituted on C1' and the acetoxy groups on C2' and C3' of the furanose rings, the copolymerization of monomers **12** and **15** with maleic anhydride yielded the alternating anhydride polymers [9–12, 14].

When cyclic vinyl ethers (**12** and **15**) were copolymerized with acrylic anhydrides in the presence of radical initiators, alternating copolymers were also produced. During copolymerization, the acrylic anhydride formed the glutaric anhydride radical at the growing chain end by cyclization, which propagated on the vinyl ether monomer. By repetition of the cross-reactions between them, the alternating copolymer was obtained [14]. Radical homopolymerization was not feasible for both monomers **12** and **15** but was feasible for acrylic anhydride. The excess feeding mole ratios of the former monomers at the onset of copolymerization were necessary for obtaining the alternating copolymers. When hydrolyzed with the aid of aqueous hydroxide, the anhydride polymers were converted to polymers (**17**, **19**, **21**, and **23**), removing the blocking groups for hydroxy and amino groups.

The number-average molecular weight (Mn) of polymer **17** measured by vapor pressure osmometry (VPO) was found to be very low (1600–1840), which corresponded to a degree of polymerization (DP) of 6 [8]. The low molecular weight of the polymer was attributed to transfer reaction on the monomer as shown in Scheme 5 and the compact structure of the polymer. The allyl and/or allyloxy radicals, formed by hydrogen transfer from the monomer to the active center, were very stable due to the formation of resonance hybrids [10]. These stable free radicals could start the copolymerization anew to result in low molecular weight. In the copolymerization of **12** with acryl anhydride, this transfer reaction was retarded owing to the longer distance between the adjacent furanose rings in the polymer chains, to yield moderate molecular weights (DP higher than 30) [14].

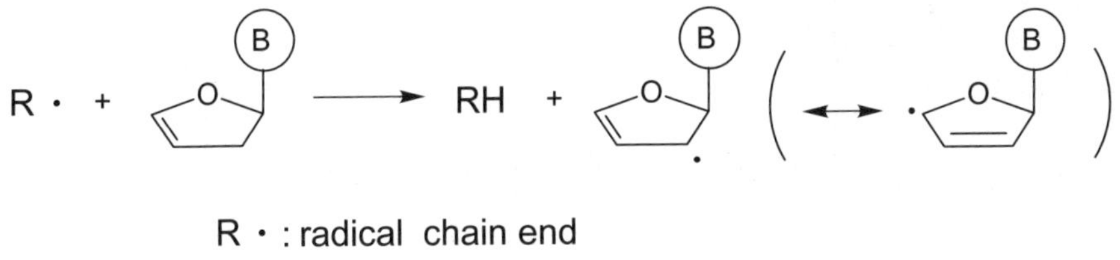

R· : radical chain end

Scheme 5

2.3.1
Physicochemical Properties

The structures of polymers **17**, **19**, **21**, and **23**, the alternating copolymers of nucleoside derivatives and dicarboxyalkylene, are quite similar to those of nucleic acids, which are the alternating copolymers of nucleosides and methylene phosphate groups. Polymers **17** and **21** are analogous to DNA whereas **19** and **23** are analogous to RNA, owing to either the absence or the presence of the hydroxyl groups on the C2' carbon atom of the furanose rings. The nucleic acid bases of the polymers were bonded to α-position of the riboses, the same as those of nucleic acids, which were believed to provide the nucleic acids with stereoregular structures. The fully extended lengths between adjacent furanose rings on the polymers (from C4' to C4") were calculated to be 3.85 Å for **17**, 4.96 Å for **19**, 6.24 Å for **21**, and 6.47 Å for **23**. The distances in polymers **17** and **19** were much shorter than that of nucleic acids (6.19 Å), whereas those of polymers **21** and **23** were very close to it. Polymer **17** had a very compact structure, so that a polymer of a high molecular weight could not be obtained. As polymer **21** had 1,3-dicarboxytetramethylene linkages replacing the methylene phosphate groups of natural nucleic acids and its backbone linked to C3' and C4', the same as that in natural nucleic acids, its structure had a close resemblance to those of natural nucleic acids. All of them

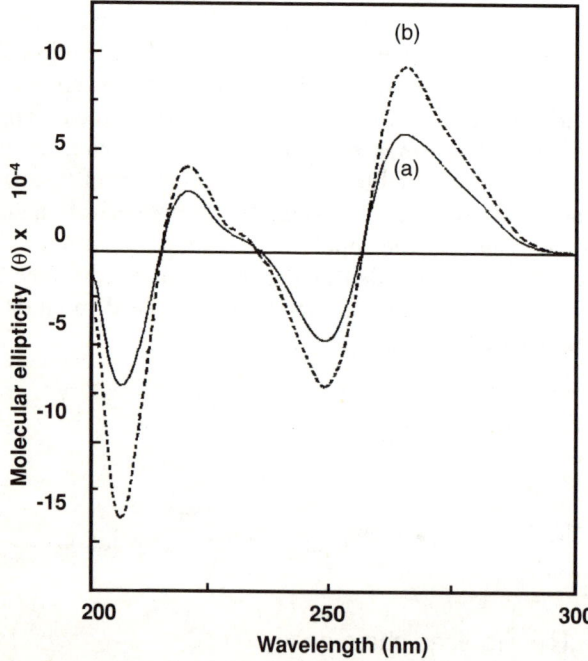

Fig. 1. CD-spectra of **23A** (a) and poly(adenylic acid) (b) in Tris-buffer (pH 7.4) at 25 °C

were soluble in water due to the hydrophilic carboxyl groups on the polymer backbones, which facilitated the investigation of their physicochemical properties in aqueous solutions similar to those of the biological systems.

The chiral atoms of C1' on deoxyribonucleosides and of C1', C2', and C3' on ribonucleosides were intact during monomer synthesis and copolymerization; polymers **17, 19, 21**, and **23** were optically active, which allowed the use of circular dichroism (CD) and optical rotary dispersion (ORD) for investigation of the polymer conformations in aqueous solutions [6, 10, 11]. The CD-curve of **23A** is shown in Fig. 1, which is quite similar to that of natural poly(adenylic acid).

The sodium salts of polymers **17, 19, 21**, and **23** were polyelectrolytes, which had viscosity as well as electrophoretic characteristics. Reduced viscosities of the sodium salt of polymer **17U**, for example, exhibited typical polyelectrolyte behavior as a function of concentration in H_2O (Fig. 2). By continuous dilution the reduced viscosities of the polymer decreased steadily and increased rapidly at the concentrations below 0.5 g/dl in water. In a neutral salt solution (NaCl, 5%) the reduced viscosity retained normal behavior.

Migration of the polyanion of polymer **17T** to an anode was investigated in an electrical field. The thin-layer electrophoresis diagram of the sodium salt of the polymer developed on an electrophoresis cellulose sheet in a buffer solution (pH=7.4) showed three separated bands (Fig. 3) [10]. Mobility of charged polymers was proportional to the net charge and inversely proportional to the

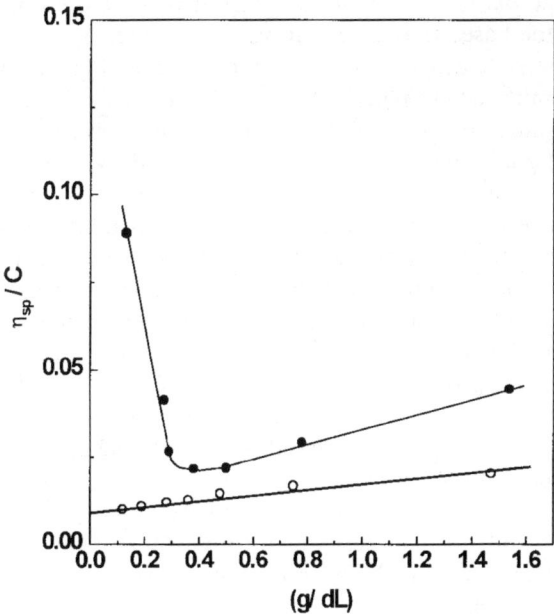

Fig. 2. Reduced viscosities vs concentration of **17U** at 30 °C in H_2O (●) and in 5% NaCl solution (○)

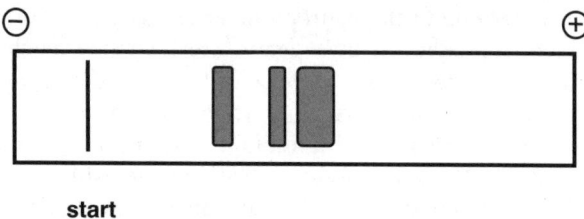

Fig. 3. Electrophoresis diagram of the sodium salt of polymer **17T** on a cellulose sheet at a constant 250 V for 2.5 h in a pH 7.4 buffer solution

two-thirds power of molecular weight [31]. The increase of each repeating unit in the polymer was accompanied by an increase of two net charges. The polymer band at a farther distance from the starting line, therefore, corresponded to the polymer of higher molecular weight. Since the average degree of polymerization of polymer **17** was found to be six by vapor pressure osmometry, the polymer should be a mixture of penta, hexa, and heptamers.

2.3.2
Base Stacking

In the biological system DNA formed a double-stranded helix, in which the sugar-phosphate backbones followed a helical path at the outer edge of the molecule and the bases were in a helical array in the central core. The bases of one strand were hydrogen-bonded to those of the other strand to form the purine-to-pyrimidine base pairs A:T and G:C. The nucleic acid bases either of the single-stranded or of double-stranded natural nucleic acids are also stacked with the adjacent bases in the same strand by hydrophobic interaction. The induced dipole-dipole interactions in the chromophores of nucleic acid bases can result in either hypochroism or hyperchroism, depending on the relative geometry of the stacked chromophores. Hypochroism is common to systems with the chromophores stacked one upon the other like a deck of cards [32, 33], while systems with chromophores in a head-to-tail aggregate are generally predicted to be hyperchromic [34]. In the measurement of UV-spectra, most of the polymers **17**, **19**, **21**, and **23** showed high hypochromicity at the maximum wavelength (250–270 nm) compared with UV-absorption of the relevant monomers as was observed in natural nucleic acids.

The UV spectra of polymer **17U** in H_2O and DMF, as a typical example, are given in Fig. 4, which showed a hyperchromicity of 30.1% in H_2O ($\varepsilon=12,360$, $\lambda_{max}=261$ nm) and a hypochromicity of 29.7% in DMF ($\varepsilon=6690$, $\lambda_{max}=268$ nm), compared with those of 2'-deoxyuridine in the relevant solvents ($\varepsilon=9500$, $\lambda_{max}=263$ nm in H_2O and $\varepsilon=9350$, $\lambda_{max}=267$ nm in DMF), respectively [11]. It is notable that hyperchromicity has rarely been found in aqueous solutions of nucleic acids though it was predicted theoretically.

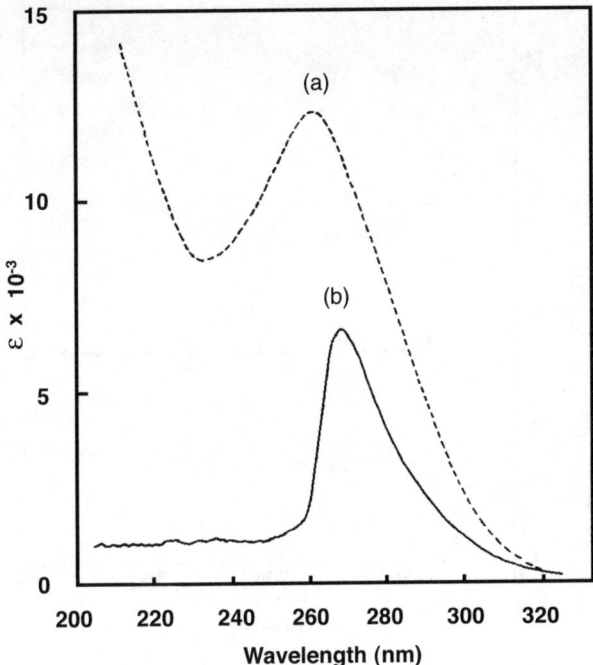

Fig. 4. UV-spectra of polymer 17U in H_2O (a) and in DMF (b)

Polymer 17U was composed of uracil-1-yl furanoid as hydrophobic and dicarboxydimethylene as hydrophilic groups. The carboxyl groups of the polymer in an aqueous solution protruded outward interacting with the aqueous environment. Consequently, the polymer had a conformation such that the chromophores aggregated into head-to-tail order in H_2O to cause the hyperchromicity. On the other hand, the uracil groups in the polymer were stacked one upon another in DMF to cause hypochromicity, which was also consistent with the fact that the UV extinction of the polymer in DMF was about half of that in H_2O.

The UV extinction and maximal wavelength of the polymer were measured in solvent mixtures of DMF-H_2O after equilibration for 24 h at 25 °C. When the water content was increased in the solvent mixture, the extinction was increased while the wavelength of absorption maxima was blue-shifted. At a solvent composition of water 50 vol %, the extinction was equal to that of free uracil as well as the mean value of the two forms.

In order to confirm whether an equilibrium state can be attained at the same solvent composition, starting from either one of the two forms, the polymer was dissolved in H_2O or in DMF and the time-dependent extinction changes were measured after diluting with DMF or H_2O, respectively (Fig. 5) [11]. The equilibrium value of extinction was attained quickly from the hypo-

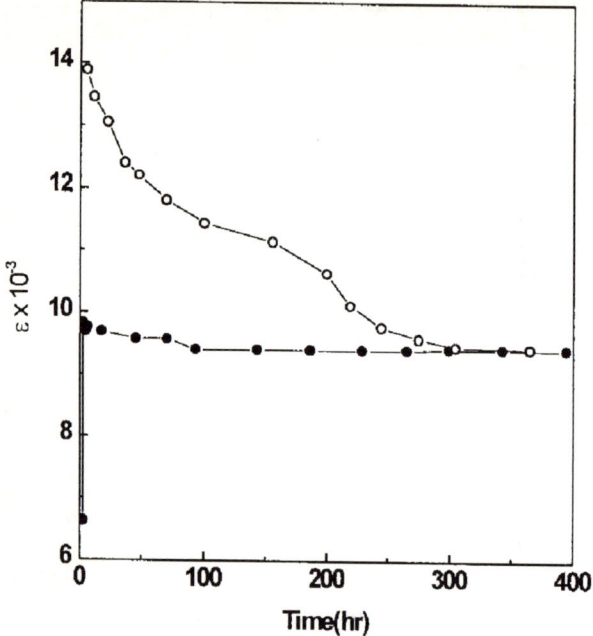

Fig. 5. Time-dependent approach to the equilibrium values of extinction coefficients (ε) at 256 nm in H_2O-DMF (1:1, v/v) at 25 °C starting with H_2O solution (●) and DMF solution (○) of polymer **17U**

chromic form although very slowly from the hyperchromic form after 270 min. The polymer **17U** contained carboxyl groups and their interactions with water, which compelled the polymer to change its conformation, seem to be stronger than those of hydrophobic groups of the polymer with the organic solvent.

Some of polymers **17** and **19** formed excimers in aqueous solution. Excimers are observed in bichromophoric molecules, where the aromatic chromophores are separated by a three-atom linkage. Excimer formation in these systems requires rotational motion around the bond of the linkage to allow two chromophores to have, within the lifetime of the excited state, a conformation suitable for complex formation in which two aromatic rings overlap in a sandwich-like arrangement [35, 36]. When these geometrical requirements are satisfied for the pendent chromophores on the polymer chains, the polymer shows an excimer fluorescence as observed in poly(vinyl aromatic)s [37] and polymers containing chromophoric pendant groups [38].

Excimer formation seemed to be favorable for polymers **17** and **19** due to the shorter distances between their nucleic acid bases, compared with those of the natural polymers. Fluorescence emissions of compound **10U** and of polymer **19U** at 20 °C, after exciting at 260 nm, were measured at the same concentrations of uracilyl groups (1×10^{-5} residue mol/l) in H_2O (Fig. 6). Compound

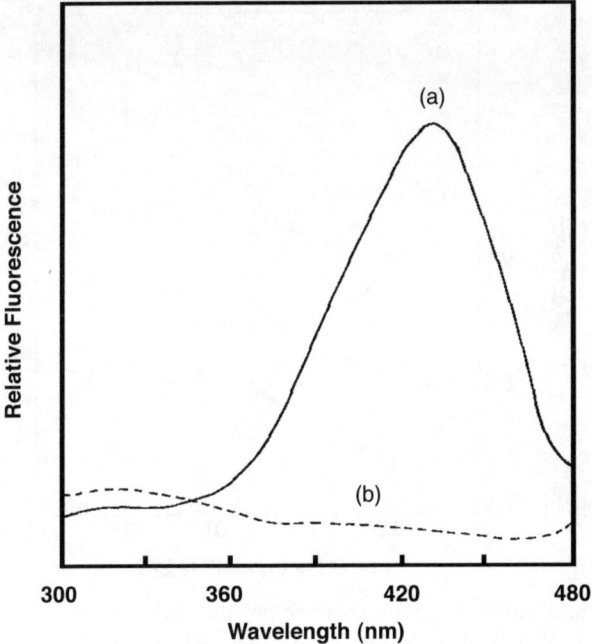

Fig. 6. Fluorescence emission spectra of polymer **19U** (a) and uridine (b) after exciting at 260 nm in H_2O at the same concentration of 1×10^{-5} residue mole/l at 25 °C

10U showed a broad peak at 335 nm with very low intensity whereas **19U** gave a broad band at 425 nm with very high intensity. The fluorescence emission of the polymer was red-shifted relative to the emission band from compound **10U**, and devoid of vibrational structures. These are typical characteristics for the excimer fluorescence, indicating that the chromophores of the polymer formed excimers in aqueous solutions. This result was further evidence for the base-stacking of the PNAs.

2.3.3
Base Pairing

The formation of complementary base-paired complexes between both RNAs and DNAs is well known. When the double-stranded complex is heated slowly in solution state, the base pairs break and finally the strands separate. This heat transition can be reversed by cooling. An intriguing question remained as to whether the synthetic PNAs interacted with natural nucleic acids to form complexes showing heat transitions. The complex formation by base-pairing and the separation into single strands are generally determined by continuous variation mixing curves [39, 40] and by the melting profile, respectively.

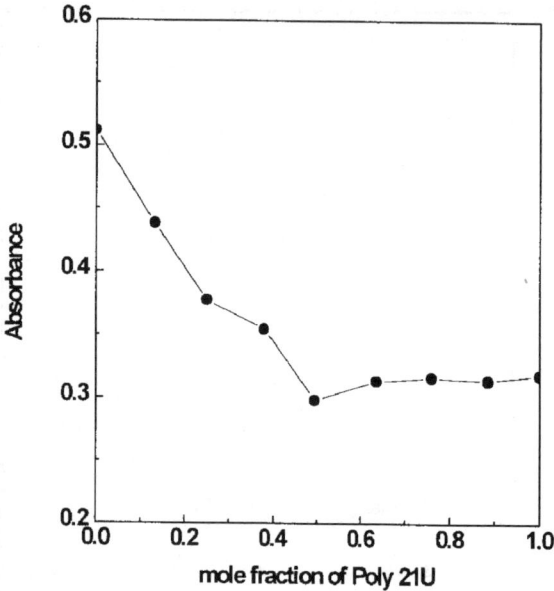

Fig. 7. Continuous mixing variation curve of polymer 21U and poly(deoxyriboadenylic acid) [poly(dA)] after three days at room temperature in Tris-buffer (pH 7.4) containing 01 mol/l Mg^{2+} and 0.1 mol/l Na^+

Base pairing of polymer 21, DNA analogues, having a distance between neighboring bases similar to those in natural nucleic acids, was extensively studied in aqueous solutions. Most of the polymers formed complexes with natural nucleic acids of complementary bases. Figure 7 shows the continuous variation mixing curves of 21U-natural poly(deoxyriboadenylic acid) [poly(dA)] of 40 bases at room temperature under biological conditions [14]. The complex was found to be a duplex of the mole ratio of 21U:poly(dA)=1:1, based on the highest hypochromicity at the same ratio. The corresponding natural complexes are well documented [41, 42].

To confirm the helical structure of 21U-poly(dA), the complex was also investigated by surface-enhanced Raman spectroscopy (SERS) [14]. As found in the studies of natural nucleic acids [43–45], the variation of the band intensities associated with the polymer backbone (sugar-phosphate and sugar-dicarboxytetramethylene) and nucleic acid base moieties was observed during the conformational change from two separated single strands to a double helical structure. As shown in Fig. 8, strong bands both of the polymer backbone and of adenine moieties were initially observed in the mixed solution of 21U and poly(dA). The band intensities associated with the adenine moieties at 730 cm^{-1} and 1330 cm^{-1} are decreased significantly in the SERS spectrum which was recorded three days after mixing, while the band intensity of 2950 cm^{-1} from the polymer backbone remained constant. This is indicative of the close proximity of the backbone on the Ag film substrate and further

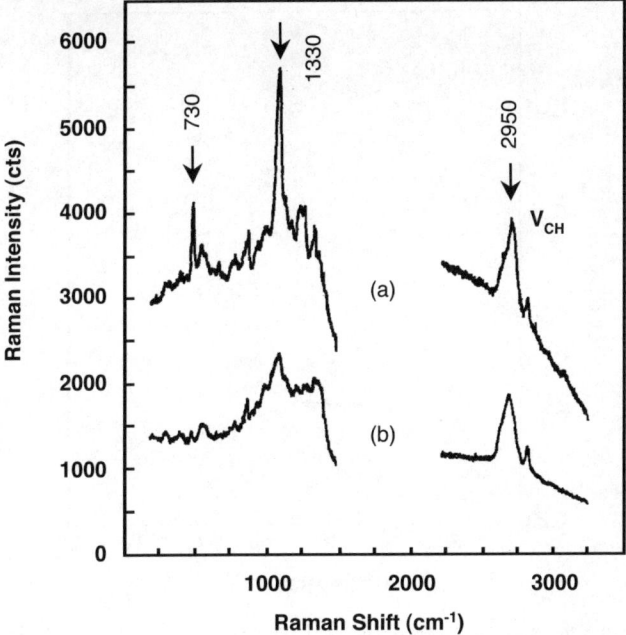

Fig. 8. SERS spectra of the mixed solution of 21U and poly(deoxyadenylic acid). The spectra were taken at 10 min (a) and 3 days (b) after mixing of the two components. Experimental condition: wavelength; 514.5 nm, power; 10 mV at the sample, integration time; 10 scans at 1 s/scan

out in the case of adenine moieties as a result of the formation of double helix by base-pairing.

Heat transition from the double helix to separate strands is accompanied by an increase of UV absorption. The melting profiles of the complexes, shown in Fig. 9, are quite similar to those of the relevant natural complexes. The melting temperature of 21U-poly(dA) was found to be 62 °C, a little lower than that (65 °C) of the natural polymer complex, poly(dA)-poly(deoxyuridylic acid) [poly(dU)] [41].

2A (Scheme 1) was also reported to form complexes with poly(uridylic acid) [poly(U)]. Due to the shorter distance between adenines in **2A** compared with that of poly(U), the complexes were proposed to be 2 poly(U)-**2A** triple-stranded, poly(U)-**2A** double-stranded, and poly(U)-25% adenine of **2A** double-stranded [46].

The RNA analogue, **23A**, formed a triplex with natural poly(uridylic acid) with the mole ratio of 1:2. While the natural triplex {poly(riboadenylic acid) [poly(rA)]-poly(ribouridylic acid) [poly(rU)]-poly(riboadenylic acid) [poly(rA)]} is transformed via a duplex [poly(rA)-poly(rU)+poly(rA)] (T_m: 62 °C) to single strands (T_m: 44 °C) [14], the complex **23A**-poly(rU)-**23A** was melted directly to the individual strands [**23A**+**23 A**+poly(rU)] at T_m of 80 °C. Com-

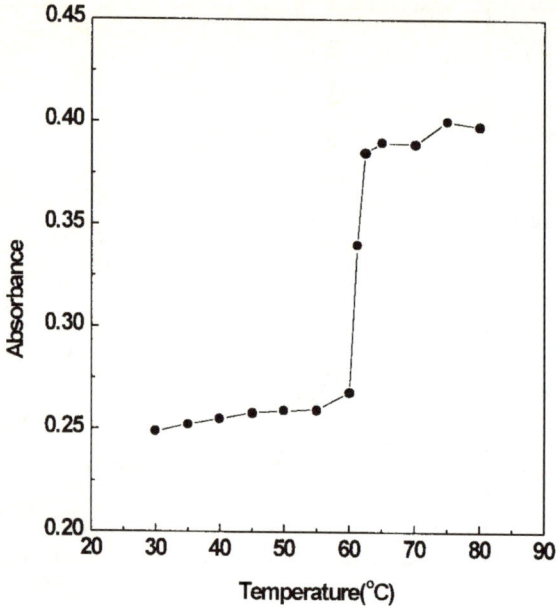

Fig. 9. Absorbance-temperature profile for 21U-poly(deoxyadenylic acid) (1:1) complex in Tris-buffer (pH 7.4) containing 0.01 mol/l Mg^{2+} and 0.1 mol/l Na^+

pared with poly(rA), **23A** contained additional C3'-OH groups, which elevated the T_m of the complex. The double-stranded complex was also formed between the synthesized RNA analogues, **23A** and **23U** with the mole ratio of 1:1, the melting point of which was found to be 64 °C. **4A** and **4U** (Scheme 1) were reported to form a stable 1:1 complex in dimethyl sulfoxide-ethylene glycol [47].

2.4
Polypeptides with Nucleic Acid Base Derivatives

Polypeptide backbones might have secondary structures as proteins do. Several alanine derivatives having nucleic acid bases were prepared and a few of them were polymerized (Scheme 6) [48–51]. Uracil and adenine were alkylated by 1-bromo-2,2-diethoxyethane and the resulting acetals were hydrolyzed and converted to β-(uracil-1-yl)-α-alanine and β-(adenine-9-yl)-α-alanine by the Strecker reaction. Thymine, guanine, and cytosine substituted alanine derivatives were also prepared in a similar manner. β-(Uracil-1-yl)-α-alanine is found as the optically pure L form in nature. Among the synthetic alanine derivatives, β-(thymine-1-yl)-α-alanine was resolved and the absolute configurations of the enantiomers were determined based on the structure of β-(uracil-1-yl)-α-alanine [50]. Step- reaction polymerizations by the active ester or mixed anhydride methods or the ring opening polymerization of N-carbox-

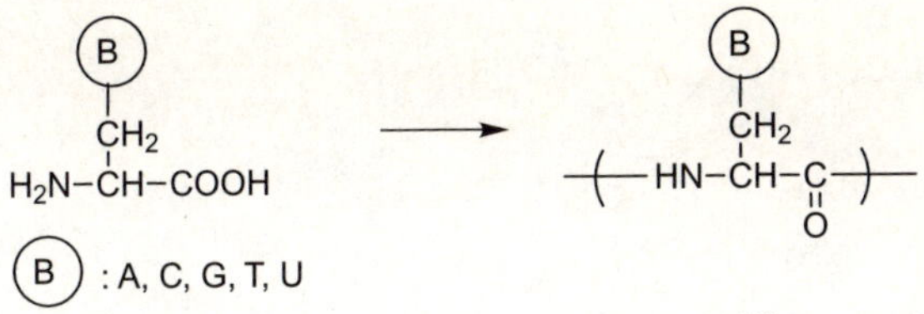

Scheme 6

B : A, C, G, T, U

24 → **25**

yanhydrides of amino acids were carried out. The polymers, however, had low molecular weights (DP<25) and showed weak interaction with the polymers containing complementary bases. Even in the case of the polymer obtained from optically pure β-(thymine-1-yl)-α-alanine, no hypochromic effect was observed on mixing with poly(adenylic acid). The polymer showed neither any evidence for base stacking, nor any secondary structures.

An interesting approach to obtain peptides in which the purine or pyrimidine residues were correctly spaced for interaction with the nucleic acids, was made by using a spacer amino acid [49]. Serine was used as a spacer and several protected tetrapeptides, *e.g.*, α-N-t-BOC-L-seryl-D,L-β-(thymin-1-yl)alanyl-L-seryl-D,L-β-(thymin-1-yl)alanine, ethyl ester, were synthesized but little was reported about their solution properties and interactions with the nucleic acids. The polymerization of the nucleic acid base-substituted L-lysine derivatives by *N*-carboxyamino acid anhydride method was also reported [52, 53].

Very recently a new family of polynucleotide analogues, so-called peptide nucleic acids (PepNAs), have been reported and have attracted a great deal of attention due to their potential uses as diagnostic and antisense agents. DNA molecules in the genes have the genetic information. The primary structure of protein, i.e., the order of amino acid residues, is specified by the sequence of the bases in the gene. The information in DNA is transcribed to the messenger RNA, and then translated to synthesize protein [54]. In drug design, inhibition of gene expression has advantages over inhibition of proteins, since a large number of *m*-RNA, and thereby proteins, are produced on transcription [55]. Moreover, the structure of proteins are usually complicated and their functions are not completely understood. Inhibition of genes can be achieved by using antisense oligonuclotides, which bind specifically with complementary nucleic acids through base pairing. This process is called hybridization. Antisense oligonucleotides have been prepared by modifying internucleotide phosphate residue in DNAs, *i.e.*, by replacing the phosphate linkers with methylphosphonates, phosphorothioates, phosphoramidates, phosphate ethers, or by using entirely different linkers such as siloxane bridges, carbonate bridges, carboxymethyl ester bridges, acetamidate bridges, carbamate bridges, or thioester bridges [55]. In addition, modified nucleoside units have also been used. The major purpose of these modifications is to increase stability against nuclease and the ability to penetrate the cell membrane.

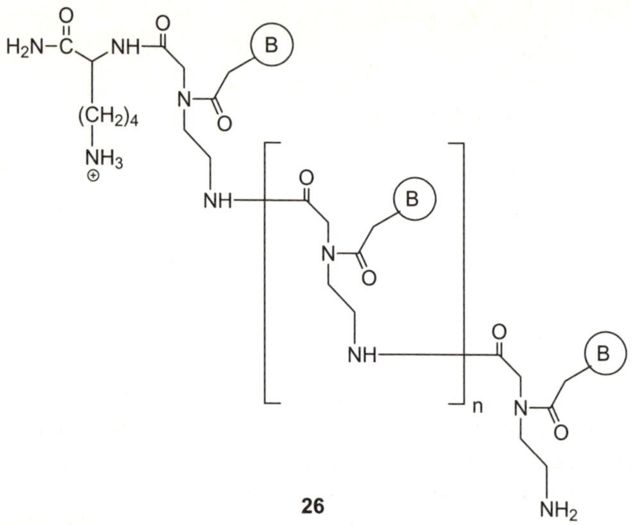

Scheme 7

In 1991, Nielsen and coworkers first reported a thymine-substituted peptide, showing remarkable stability against enzymatic degradation and also outstanding hybridization properties with DNA (Scheme 7) [56]. In a computer model, the distance between nucleobases suitable for hybridization with DNA was estimated. The optimal number of bonds between the nucleobases was found to be six and the optimal number of bonds between the backbone and nucleobase to be two to three. To meet these structural requirements, (2-aminoethyl)glycine was used as a repeating unit of the backbone and nucleobases were attached to the backbone *via* methylenecarbonyl linkers [56–63]. Oligonucleotides with ten nuclobases were synthesized by Merrifield's solid phase method [57]. The lysine was attached to the C-terminal to diminish self-aggregation of the oligonucleotides. As complementary DNAs, $(dA)_6$, $(dA)_8$, and $(dA)_{10}$ were used. The PepNAs formed complexes with complementary DNAs. They showed well-defined melting curves and the melting temperature (T_m) increased as the PepNA length was increased. T_m values were much higher than those of the corresponding DNA/DNA complexes. This was attributed to the lack of electrostatic repulsion between PepNA and DNA and the constrained flexibility of the polyamide backbone. T_m values for the complexes of DNAs containing mismatches such as $(dA)_2(dG)(dA)_2(dG)(dA)_4$ were lower, indicating formation of specific hydrogen bondings between complementary bases.

In circular dichroism study on the binding of an octamer of PepNA having thymine and a terminal lysine, a right-handed triplex was found to form between the PepNA and poly(dA) [60]. Since PepNA was achiral except for the terminal lysine, it showed a very weak CD spectrum. When 1 mole of poly(dA) was mixed with 2 moles of the PepNA, the CD spectrum of poly(dA) was changed dramatically, indicating the formation of a triplex. A remarkable thermal stability of the triplex was also observed in the same study. The CD spectrum of poly(dA)[PepNA-T_8]$_2$ triplex was very similar to that of poly(dA)[poly(dT)]$_2$ triplex. The PepNA-PepNA duplex was also reported. Two complementary PepNA decamers, H-GTAGATCACT-L-Lys-NH$_2$ and H-AGTGATC-TAC-L-Lys-NH$_2$, were prepared [61]. Neither PepNA decamer showed any strong CD spectrum resulting from helicity. However, when two complementary PepNAs were mixed in a 1:1 stoichiometric ratio, a strong CD spectrum was observed. Very interestingly, the mirror-image CD spectrum was obtained when the terminal L-lysine was substituted with D-lysine. Further study showed that the preferred chirality of a PepNA-PepNA duplex was very dependent on the nucleobase next to the chiral terminal amino acid [62]. Only if guanine or cytosine was the base next to the terminal amino acid did a duplex of a stable helical structure form. The preferred helical sense was determined by the absolute configuration as well as the nature of the side chain of the terminal amino acid. The hydrophobic side chains induced right-handed helices. A PepNA-PepNA-PepNA triplex was also reported to form between one adenine-PepNA decamer and two thymine-PepNA decamers [63].

Since the first report on the peptide nucleic acids by Neilson, numerous reports on the synthesis of amino acids having nucleic acid bases and their polymerization have appeared. PepNAs have amide groups on their backbones which can exist in *E*- or *Z*-form. It was found that all the amide bonds between nucleobases and the polymer backbones were oriented in *Z*-form in PepNA-DNA or PepNA-PepNA complexes, while the monomers of PepNAs or single strand PepNAs had both *Z*- and *E*-isomeric units in solution. In order to control the stereochemistry of these amide groups, monomer 27 was synthesized, where the central amide bond of PepNA was replaced by a configurationally defined C-C double bond (Scheme 8) [64]. The authors expected that the PepNAs derived from monomer 27 would show the similar hybridization ability with natural nucleic acids, since the deleted carbonyl group is not involved in any hydrogen bondings which would result in complexation.

PepNAs are achiral and insensitive to CD, but optically pure polyamide nucleic acid analogues were also reported. Monomer 28 (Scheme 8) has naturally occurring amino acid units, which will confer chirality on the resulting polymers [65]. Although stereocenters on the backbones were expected to influence the complexation of the PepNAs, no details were reported. Polymer 29

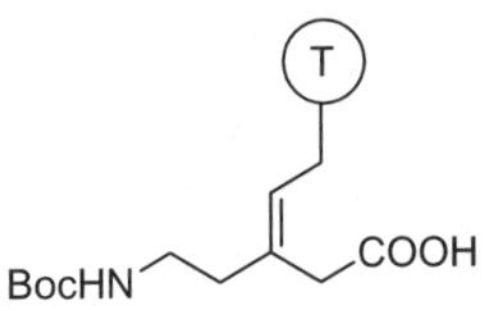

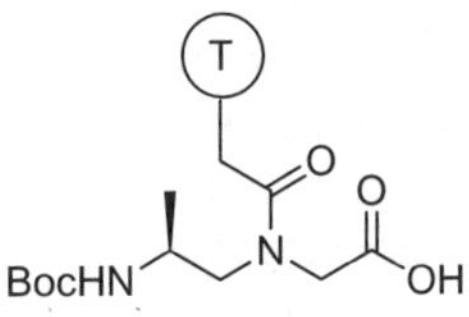

Scheme 8

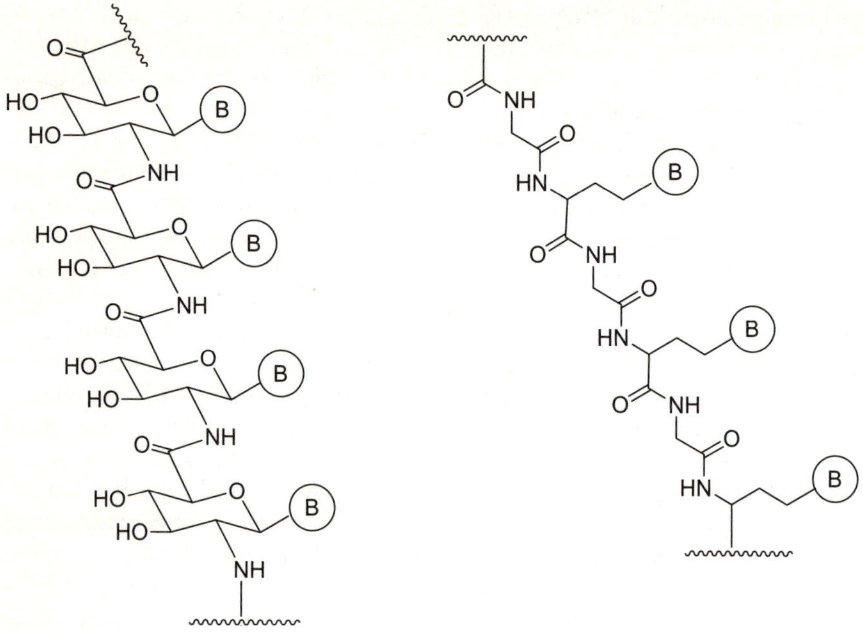

29   **30**

Scheme 9

in Scheme 9 has a glucopyranosyl amide backbone. It was soluble in water and showed hybridization with DNA [66, 67].

Polymer **30** was named α-PepNA. Nucleobases were attached to α-position *via* ethylene linkages of L-α-amino acids. The amino acids were coupled with glycine to yield optically pure monomers of α-PepNA [68].

3
Alteration of Polynucleotide Analogues

3.1
Dimerization of Nucleobases by UV Irradiation

Photochemical reactions of purines and pyrimidines in light are well known, and lead to mutations. Although various photochemical reactions can occur with both pyrimidines and purines, photodimerization of thymine is of the utmost importance. Photochemical reaction by UV irradiation of thymine results in formation of dimers with cyclobutane rings, which block DNA replication. Photodimerization was reported with the synthetic polymers having pendant thymine bases. In a study with polyacrylates and polymethacrylates with thymine and their dimer model compounds in the solvents, DMSO, DMF, and a DMSO-ethylene glycol (EG) mixture, quantum efficiencies for photodimerization for the polymers about ten times higher than those for the dimer model compounds were found (Scheme 10) [69]. The quantum efficiencies for the dimers were almost invariant with the solvents. However, the quantum efficiencies for the polymers increased in the order of DMSO-EG>DMF>DMSO. This result was related to the self-association by base pairing. Because the polymer was soluble in DMSO and insoluble in EG, the polymer chain had the expanded conformation in DMSO, while the self-association of thymine bases occurred in EG, resulting in compact structures.

The results obtained with poly-L-lysine containing pendant thymine were especially interesting [70]. In the study of the 88% thymine grafted poly(L-lysine) by CD and UV at pH above 10, the helicity of the polymer was found to increase by photodimerization of thymine up to a 30% conversion, and then decrease. After photodimerization, the helical structure was not affected by the pH change, indicating that the reaction occurred between thymine bases on the same backbones, resulting in fixation of the structure. Decrease of the helicity at high conversion was ascribed to distortion of the helical structure by a reaction between remote grafted thymine bases.

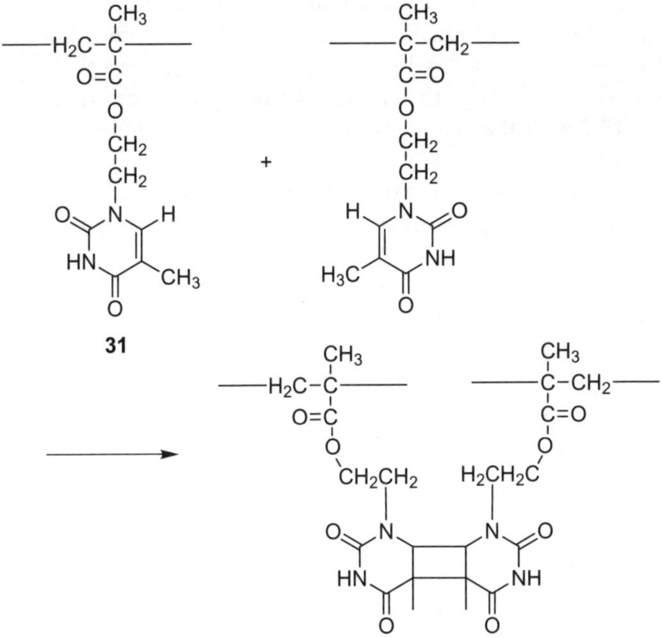

Scheme 10

3.2
Depurination and Depyrimidination

The depurination or depyrimidination of nucleic acids, the release of purine or pyrimidine bases, respectively, from nucleic acids by hydrolysis of the *N*-glycosidic bonds, gives rise to alterations of cell genome [71, 72]. The rate constant of depurination under biological conditions was found to be 3×10^{-11} s^{-1}, *i.e.*, a mammalian cell containing 3.86×10^9 nucleotides spontaneously loses 10,000 bases from its genome in 24 h [73]. The rate is increased as the pH is lowered or as the temperature is elevated. The apurinic sites resulting from depurination are quite stable [74], and the cell evolves mechanisms to repair these lesions [75]. Unrepaired apurinic sites are to have two biological consequences – lethality [76, 77] and base-substitution errors [77].

Depurination and depyrimidination were found to occur in polynucleotide analogues with much increased rates. When polymers **17** and **19** were dissolved in a buffer solution (pH 7.4) above 30 °C, the *N*-glycosidic bonds of the polymers were spontaneously hydrolyzed to liberate nucleic acid bases from the polymer backbones, as shown, *e.g.*, in Scheme 11 for polymer **19**. Such a reaction was not observed in polymers **21** and **23**.

The rates of deurination of **19H** [78], for example, at different temperatures are shown in Fig. 10. The reaction at the beginning was very fast at a higher temperature (90 °C) and slow at a lower temperature (37 °C). To determine the rate constants, the logarithmic concentrations of the hypoxanthine remaining on the polymer chain were plotted against time (Fig. 11), which obeyed first-order kinetics up to 6 h of reaction time. The initial rate constant at 37 °C was measured to be 1.9×10^{-6} s^{-1}, which is 10^5-fold higher than that (3×10^{-11} s^{-1}) [73] of natural DNAs under the same conditions. Depyrimidination was also observed in **19U** with the rate constant of 2.5×10^{-4} s^{-1} under the same conditions.

Since a similar reaction did not occur with polymers **21** and **23** under the same conditions, depurination and depyrimidination can be explained by

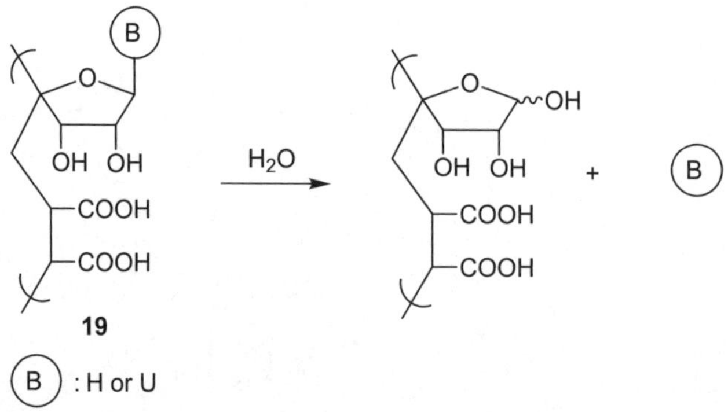

Scheme 11

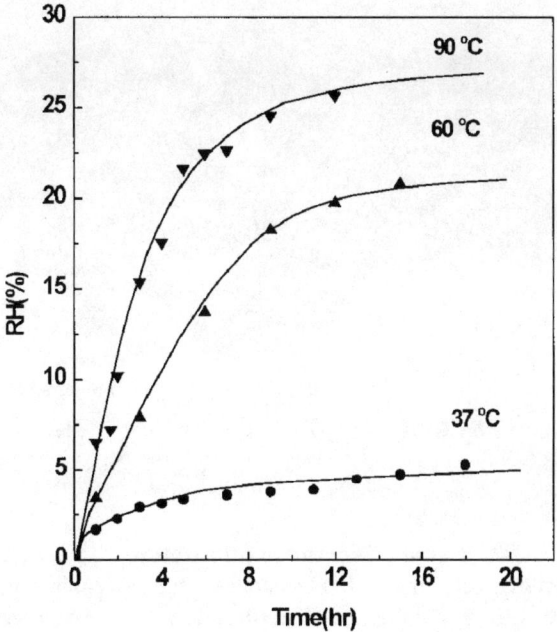

Fig. 10. Released hypoxanthine from polymer **19H** as a function of time at pH 7.4 and different temperatures

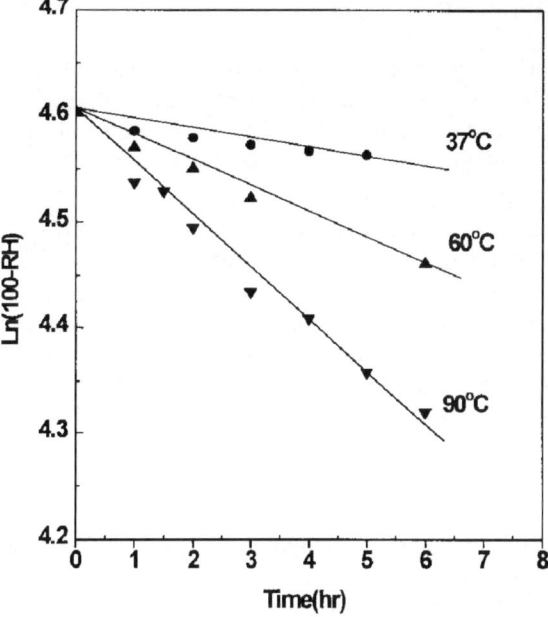

Fig. 11. Logarithmic concentrations of hypoxanthine remaining on the polymer **19H** [ln (100–RH)] *vs* time at pH 7.4 and different temperatures

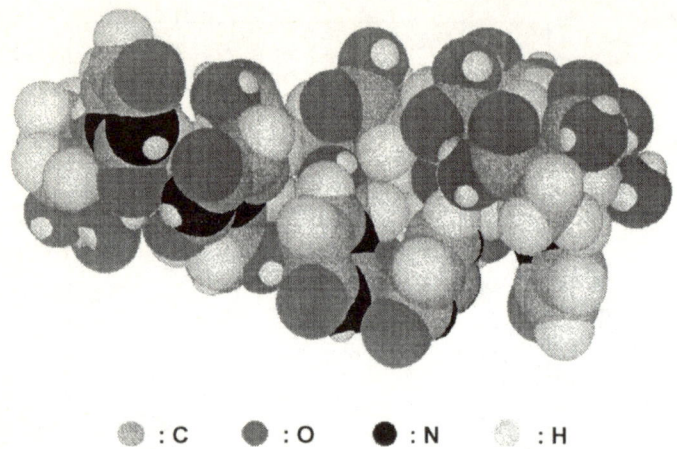

● : C   ● : O   ● : N   ● : H

Scheme 12

steric effects. Using a simple energy minimization calculation, one of the probable conformations of the lowest energy for polymer **19U** is shown in Scheme 12. It has a very compact structure. The fully extended distance between adjacent ribose rings (from C4' to C4") of polymer **19** was 4.96 Å, which was much shorter than that (6.19 Å) of the corresponding nucleic acid. This made environment of the bases so crowded that it forced a definite amount of them to be substituted by small groups (OH) to release the strain of the polymer chains.

The activation energy of the depurination of **19H** was found to be 10 kcal/mol by an Arrhenius plot, about 20 kcal/mol lower than those (31–34 kcal/mol) [79–81] of the depurination of DNAs in vitro and of hydrolysis of pyrimidines of deoxyribonucleosides. Due to the crowded environment around the hypoxanthine groups, the ground state potential energy of the N-glycosidic bond in the polymer was elevated more than that of the transition state, and thus the activation energy for the reaction decreased and the hydrolysis accelerated.

Since natural nucleic acids often have compact structures due to the crowded environment around the bases created by the intricate chain folding, depurination or depyrimidination may be accelerated in a similar manner in the biological system.

4
Catalytic Activity for Phosphodiester Hydrolysis and DNA Cleavage

RNA (ribozyme) showed enzymatic activity for transphosphorylation or hydrolysis of natural nucleic acids [82–85]. In the self-splicing of r- or m-RNA precursors the cleavage and ligation of the strands occurred to remove the

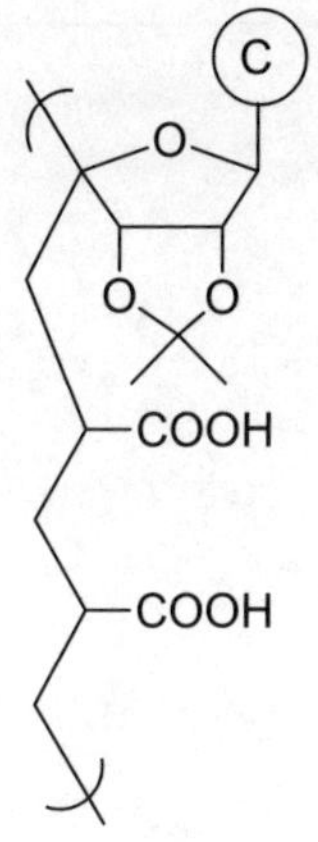

32

Scheme 13

intron so that matured RNAs resulted [86–91]. The precursor to t-RNA^{tyr} was matured by hydrolysis with the aid of M1 RNA [92–95]. Moreover, DNA cleavage was also catalyzed by the ribozyme from *Tetrahymenena* [96, 97]. Although the action mechanism of ribozymes is still obscure, it is probable that 2',3'-diol groups of furanose rings play key roles.

Polymers **23A** and **23C** are analogous to RNAs. Very interestingly, they showed catalytic activities for the hydrolysis of phosphodiesters and the cleavage of nucleic acids. To elucidate the action mechanism of the hydrolysis, the acetonide (**32**) of polymer **23C** was synthesized (Scheme 13).

Hydrolysis rates of p-nitrophenyl phosphate substrate were measured in Tris buffer (pH 7.4) with the ultraviolet absorption (400 nm) of the p-nitrophenol evolved. Figure 12 shows the hydrolysis rate of the substrate in the buffer solution, both in the presence and in the absence of the polymers. The hydrolysis rate was enormously accelerated by polymer **23C**, while no catalytic activity was found in the presence of polymer **32**. It showed clearly that *vic-cis*-diol groups of furanose rings played key roles in this catalysis, since the activity was lost when the diol groups were blocked by isopropylidene groups (**32**).

The initial catalysis rates were obtained by fixing concentration of polymer **23C** and changing the substrate concentration. Michaelis-Meten kinetics for the polymer was confirmed by plotting the double reciprocal form of Lineweaver and Burk ($1/v$ vs $1/s$) (Fig. 13), which gave K_m and V_{max}. The k_{cat} was obtained from V_{max} ($k_{cat}=V_{max}/[E]_o$), assuming that one polymer molecule had one active center, since the polymer with Mn of lower than 4000 showed no significant catalytic activities. k_{cat} was found to be $6.0\times10^{-1}\,h^{-1}$ for **23C**, which was 670-fold higher than the $9.12\times10^{-4}\,h^{-1}$ of the uncatalyzed reaction [98].

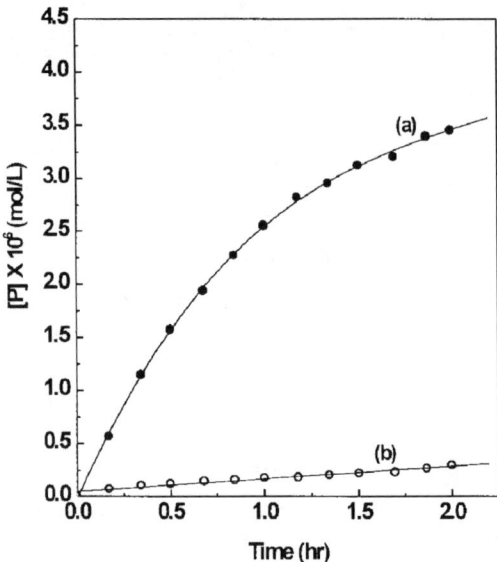

Fig. 12. Concentrations of *p*-nitrophenol evolved [P] during hydrolysis of ethyl *p*-nitrophenyl phosphate as a function of time in the presence of polymers **23C** (a) and **32** (b) in Tris buffer (pH 7.4), 50 °C, and ionic strength=0.02 (KCl). [polymer]=9.61×10^{-6} mol/l, [substrate]=4.67×10^{-4} mol/l

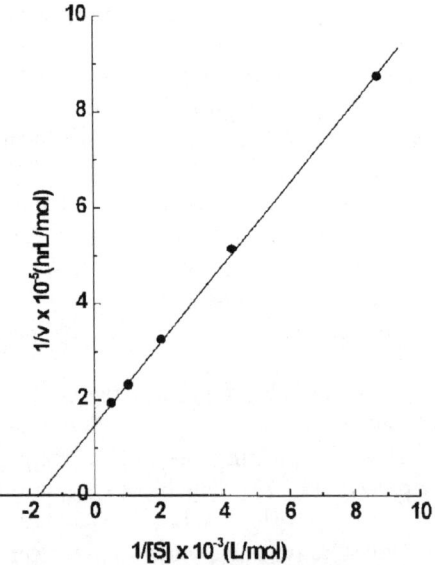

Fig. 13. Reciprocals of the initial rates as a function of the reciprocals of the substrate concentration at a constant concentration of polymer **23C**. [polymer]=9.29×10^{-6} mol/l in Tris-buffer (pH 7.4) at ionic strength of 0.02 (KCl) and 50 °C

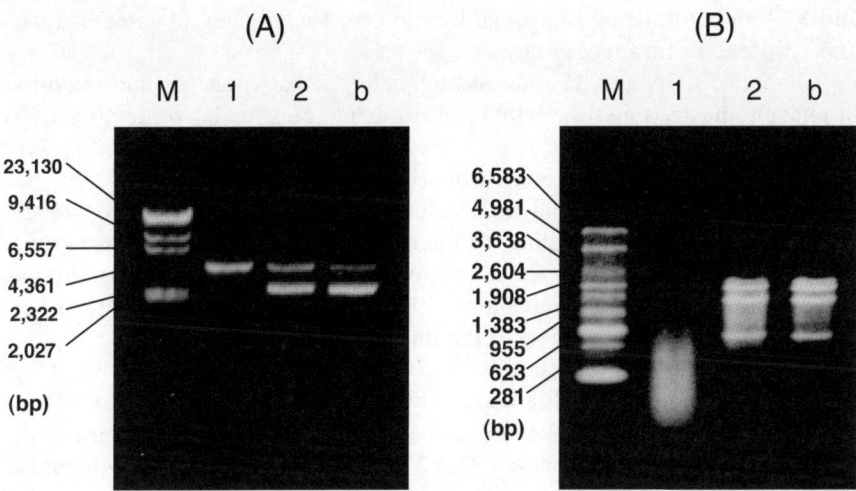

Fig. 14 A,B. Electrophoresis diagram of the reaction mixtures on the agarose gel: nucleic acids: **A** DNA; **B** RNA – both were incubated at 37 °C, pH=7.4 (Tris-buffer), and ionic strength of 0.02 (KCl) for 6 h (lane 2), in the presence of polymer **23C** (lane 1), lane b; DNA or RNA, lane M; Maker DNA; pBR322, RNA; BMV. [polymer]=10.7 pmol/20 μl [DNA]=60 ng/20 μl, and [RNA]=25 ng/20 μl

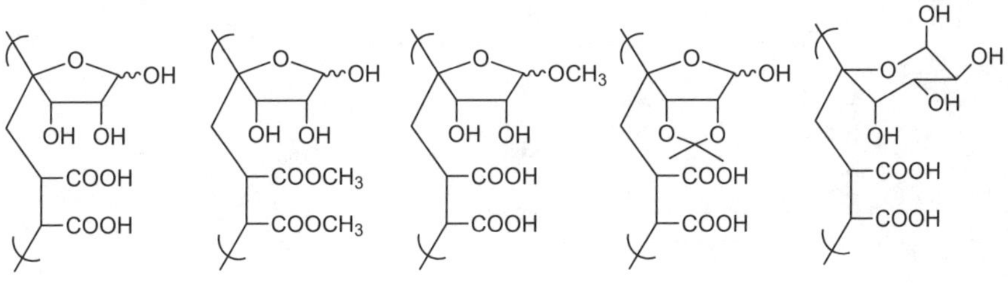

Scheme 14

Supercoiled double-stranded DNA and RNA were incubated in the presence of polymers **23C** and **32** at pH 7.4 and 37 °C and developed on agarose gel (Fig. 14). Polymer **23C** catalyzed the cleavage of DNA to result in band II after 6 h, while no cleavage was observed with **32**.

From these results it was concluded that the catalytic activity was caused by the *vic-cis*-diols of ribose rings but not by the nucleic acid base (*e.g.*, cytosine), which might play a role in the formation of active centers. To elucidate the action mechanism of the catalysis, five polymeric model compounds without nuclei acid bases were synthesized by copolymerizing sugar derivatives with maleic anhydride, and subsequent hydrolysis (Scheme 14) [99, 100]. Starting from polymer **33**, the carboxyl groups and 1'-OH groups were blocked by methylation and acetylation to result in **34** and **35**, respectively, and 2',3'-*vic-cis*-diols by isopropylidene to give **36**. Polymer **37** contained pyranose instead of furanose rings.

Polymers **33**, **34**, and **35** showed high catalytic activities for the hydrolysis of phosphodiester substrate (ethyl *p*-nitrophenyl phosphate) while no activity was observed for **36** and **37**. The k_{cat} were found to be 0.42, 0.90, and 0.17 h^{-1} for **33**, **34**, and **35**, respectively, which were about 10^3-fold faster than that ($9.12 \times 10^{-4} h^{-1}$) of the uncatalyzed reaction. These results showed that *vic-cis*-diols of ribose were responsible for the catalysis and that the carboxyl groups played no significant roles. The activation energy of the catalysis for polymer **33** was found by an Arrhenius plot to be 6.9 kcal/mol, which was 6.6 kcal/mol lower than that (13.5 kcal/mol) of the uncatalyzed reaction.

Competitive inhibition was observed in the catalysis of polymer **33** on addition of acetate ions and the dissociation constant of the inhibitor-polymer complex was found to be 5.9×10^{-4} mol/l. By addition of K_2HPO_4, the noncompetitive inhibition was observed ($K_I = 2.5 \times 10^{-4}$ mol/l) (Fig. 15). Both inhibitions occurred by blocking *vic-cis*-diols through the formation of hydrogen bondings. The acetate ion competed with the substrate for forming H-bonds with the polymeric catalyst. The phosphate ions, as dianions, were stronger H-bond acceptors with *vic-cis*-diol groups of the polymer catalyst than was the substrate, leading to a noncompetitive inhibition.

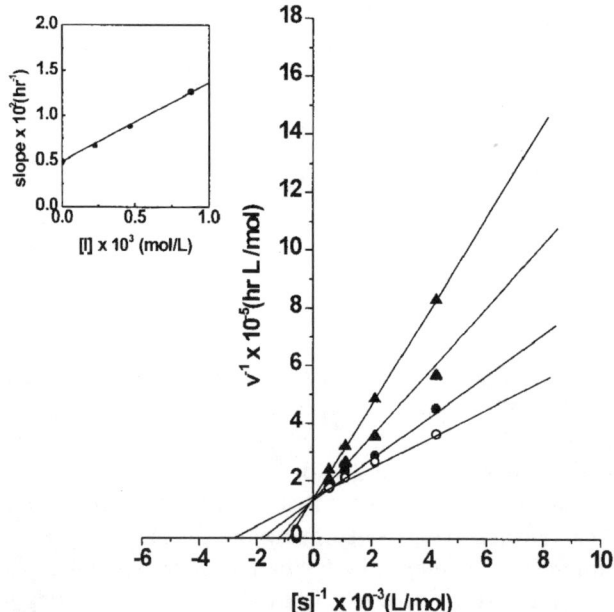

Fig. 15. Double-reciprocal plots ($1/v$ vs $1/[S]$) measured at different concentrations of the inhibitor, sodium acetate, in the presence of polymer **33** in Tris-buffer (pH 7.4) and at ionic strength of 0.02 (KCl) and 50 °C. [polymer 33]=1.62×10^{-5} mol/l. K_I determination is shown in the *upper left plot*

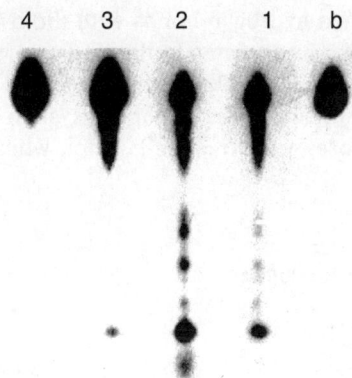

Fig. 16. Autoradiodiagram of 7 mol/l urea – 8% acrylamide gel electrophoretic analysis of the reaction mixture. ^{32}P-labeled ssDNA of 30 bases was incubated at 37 °C, ionic strength=0.02 (KCl), pH=7.4 (Tris-buffer) for 12 h (lane b), in the presence of polymers **33** (lane 1), **34** (lane 2), **35** (lane 3), and **36** (lane 4). [DNA]=7.5×10^{-4} mol/l, [polymer]=7.3×10^{-5} mol/l

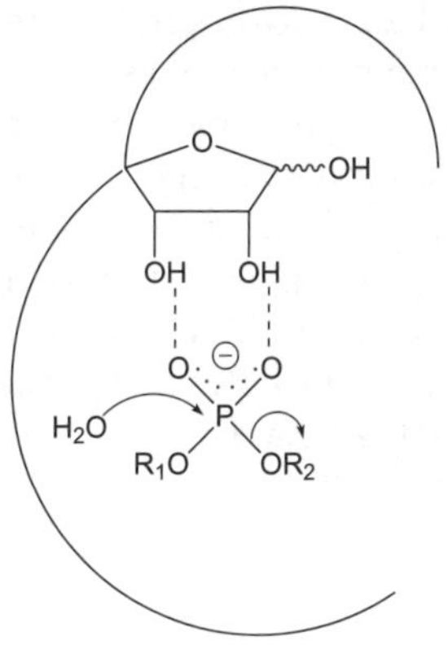

Scheme 15

^{32}P-labeled on 5' of ss DNA of 30 bases, (CATGGCAAAGCCAGTATACA AATTGTAATA), corresponding to the human foamy virus proviral DNA in the position of nucleotide 3643–3611, was incubated in a Tris buffer (pH 7.4) at 37 °C and ionic strength 0.02 (KCl) in the presence of polymers **33–36**. Autoradiograms of acrylamide gel electrophoretic analysis of the reaction mixtures are shown in Fig. 16. No DNA cleavage occurred in the buffer or in the presence of **36**, while polymers **33–35** catalyzed hydrolysis of DNA, as was observed in the hydrolysis of ethyl *p*-nitrophenyl phosphate substrate [99, 100].

As the catalytic activity was not found in the presence of α-D-ribose itself, the polymer backbone likely formed active centers by chain folding. The ribose rings having *vic-cis*-diol groups were located inside the active sites, where the phosphodiester substrates were also accommodated (Scheme 15).

The *vic-cis*-diol groups formed hydrogen bonds with the two phosphoryl oxygen atoms of the substrate so as to activate the phosphorus atoms to be attacked by nucleophiles (H_2O) and the alkoxy groups to be left either by an in-line or by an adjacent mechanism. In the biological system there are plenty of biopolymers containing riboses with free *vic-cis*-diol, which might show nuclease activities.

5
Application of Polynucleotide Analogues

5.1
Liquid Chromatography

Specific base pairing with complementary bases through hydrogen bondings was observed in many synthetic polynucleotides. This complementary base recognition ability of polynucleotides can be applied to high performance liquid chromatography for separation of oligonucleotides. Nucleic acid derivatives were immobilized on 3-aminopropylsilanized silica gel and used as a packing material [101]. To reduce a hydrophobic interaction of the mixture with the stationary phase, methanol was used as an eluent. The chromatogram of nucleoside mixtures showed that the complementary nucleosides eluted later than the other nucleosides, indicating the specific hydrogen binding between the bases on the stationary phase and the complementary nucleosides. In another study, poly(L-lysine) having pendant nucleic acid bases was immobilized on silica gel and employed in packing the columns [102]. Oligoethylenimine derivatives having nucleic acid bases were separated successfully with the column in the methanol mobile phase.

5.2
Template Polymerization

The specific interactions between complementary bases were also applied to template polymerization [103]. Three paths for the template polymerization of a monomer having uracil in the presence of the template having adenine are possible; in path (a) the polymerization takes place after base pairing of the monomers with the bases on the template. If the interaction between the monomers and the template is not so strong, the complex of the template will be formed only with growing oligomers (b) or the polymer (c). In the case of path (a), the rate of the initiation, propagation, and termination will be dependent on the presence of the template, in the case of (b) the rate of termination will be retarded to cause the acceleration of the polymerization rate, and in the case of (c) the rate of polymerization will be independent of the presence of the template. It was found that the polymerization of the methacrylate monomer having a nucleic acid base followed path (b) in the presence of the polymethacrylate containing nucleic acid bases. Copolymerization of **38** with **40** in

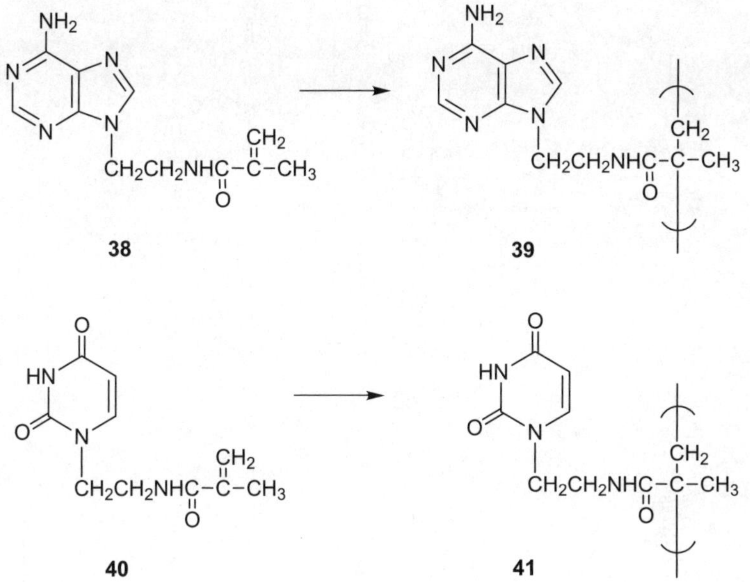

Scheme 16

the presence of template polymer **39** was carried out in DMSO-EG (3/2, v/v) at 60 °C with AIBN (Scheme 16). During polymerization the polymer precipitated, and this precipitated polymer was identified as a stable **39–41** polymer complex. The polymer remaining in the solution was **39**, which was used as the template polymer. This result shows that the template polymerization of **40** occurred in the presence of polymer **39**, where **40** was bound to **39** *via* specific hydrogen bonding before polymerization but **38** was not.

5.3
Photoresist

Photodimerization of pyrimidine bases is a reversible reaction. They undergo photodimerization by exposure to UV light above 270 nm, and readily reverse to the monomers by UV light at shorter wavelengths. The reaction was applied to the preparation of deep-UV negative and positive-type photoresists [104–106]. The photochemical reaction of poly(methacrylate)s having thymine derivatives was studied in solution and in the film state. The quantum yields of the polymers in the film state were found to be 5–9 times higher than those in solution, showing the possibility of intermolecular reaction increases [105]. However, the intramolecular reaction was also expected to occur more readily in the film state, which was unfavorable for high resolution of the photoresists. The photoreversal reaction of the pyrimidine photodimers was studied with

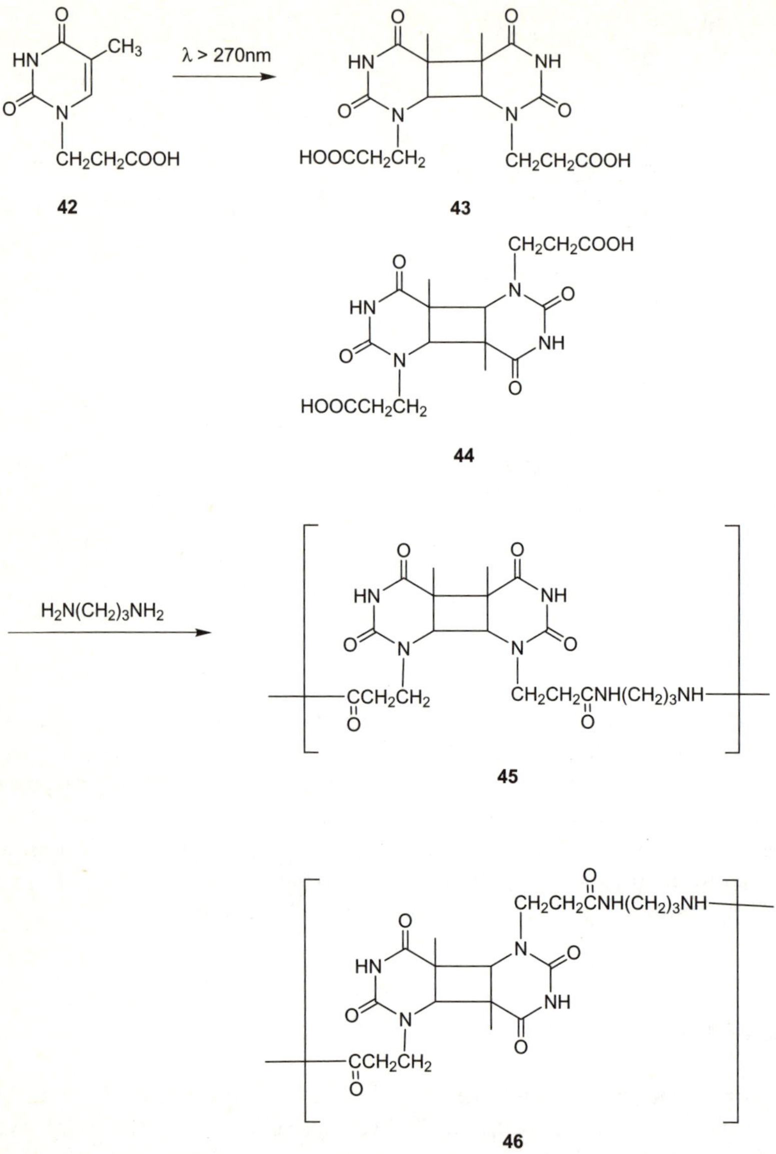

Scheme 17

the polymers containing pendent thymine photodimer units [106]. Monomeric photodimers **43** and **44** were synthesized by a photochemical reaction of thymine derivative **42**. Under the reaction conditions, four isomers formed and the *cis*-isomers were isolated for polymerization (Scheme 17). Polyamides were obtained by polycondensation of the diacid monomers with 1,3-diaminopropane *via* the activated ester method. The irradiation on the polymer film at 254 nm caused photodissociation of the thymine dimers, resulting in the polymer chain breakage. The application of the polymers as positive-type photoresists was reported to show resolution of over 0.5 µm.

6
Future Perspectives

It is the most striking feature of PNAs that they can recognize the molecules of particular structures. The specific bindings of PNAs have been observed with natural nucleic acids or other PNAs of complementary bases through hydrogen bondings. The binding patterns have become more elaborated by recent synthetic efforts to make PNAs more closely resemble natural nucleic acids. As certain PNAs form the double-stranded helical complexes with the natural nucleic acids of complementary bases showing denaturation at high temperatures, they can be used as templates for the polymerase chain reaction (PCR). The polymer backbones consist of C-C bonds which are not susceptible to hydrolysis and thus the templates can be used repeatedly. Since the solubilities of the synthetic polymers are quite different from those of natural nucleic acids, the separation of the new polymers obtained by PCR can be accomplished by a simple process. Two complementary strands of synthetic homopolyribonucleotides, such as poly(I)-poly(C) or poly(A)-poly(U), have been extensively studied as interferon inducers [107, 108]. The former also inhibited the rate of growth of human tumor xenografts in mice and even caused tumor regression. These activities were deteriorated by cleavage of the chains through hydrolysis in the biological system. Owing to the resistivity of the C-C bonds to hydrolysis, the double-stranded RNA analogues have potential uses as immunomodulating chemotherapeutic agents. When the sequences of bases can be controlled, they could be used as antitemplate compounds by blocking the functions of templates or primers, as antisense compounds by blocking m-RNAs translations, and as antigene compounds by triplex formations. Peptide nucleic acids have already attracted great attention as antisense agents. Finally, it will be noteworthy to see whether the ribozyme analogues can be obtained by introducing bases to polymers 33 or 34, which might cut the target RNA at predictable positions for the chemotherapy of tumors or viral diseases.

References

1. Watson JD, Crick FHC (1953) Nature 171:737
2. Inaki Y (1992) Prog Polym Sci 17:515
3. Takemoto K, Inaki Y (1987) In: Takemoto K, Inaki Y, Ottenbrite RM (eds) Functional monomers and polymers. Marcel Dekker, New York, p 149
4. Overberger CG, Inaki Y (1979) J Polym Sci Chem Ed 17:1739
5. Overberger CG, Inaki Y, Nambu Y (1979) J Polym Sci Chem Ed 17:1759
6. Overberger CG, Chang JY (1989) J Polym Sci Chem Ed 27:3589

7. Han MJ, Cho TJ, Chang JY (1996) Polynucleotide analogues. In: Salamone JC (ed) Polymeric materials encyclopedia. CRC, Boca Raton, vol 8, p 6402
8. Han MJ, Park SM (1990) Macromolecules 23:5295
9. Han MJ, Lee CW, Kim KH, Lee SH (1992) Macromolecules 25:3528
10. Han MJ, Park SM, Park JY, Yoon SH (1992) Macromolecules 25:3534
11. Han MJ, Chang YS, Park JY, Kim KH (1992) Macromolecules 25:6574
12. Han MJ, Kim KS, Cho TJ, Kim KH, Chang JY (1994) Macromolecules 27:2896
13. Han MJ, Park SW, Cho TJ, Chang JY (1996) Polymer 37:667
14. Han MJ, Lee GH, Cho TJ, Park SK, Kim JH, Chang JY (1997) Macromolecules 30:1218
15. Hafird MH, Jones AS (1968) J Chem Soc (c) 2667
16. Nielsen PE, Haaima G (1997) Chem Soc Rev 73
17. Overberger CG, Morishima Y (1980) J Polym Sci Polym Chem Ed 18:1247
18. Ludwick AG, Overberger CG (1982) J Polym Sci Polym Chem Ed 20:2123
19. Chu VP, Overberger CG (1986) J Polym Sci Polym Chem Ed 24:2657
20. Lan M-J, Overberger CG (1987) J Polym Sci Polym Chem Ed 25:1909
21. Overberger CG, Kikyotani S (1983) J Polym Sci Polym Chem Ed 21:525
22. Ishikawa T, Inaki Y, Takemoto K (1978) Polym Bull 1:85
23. Anand N, Murthy NSRK, Naider F, Goodman M (1971) Macromolecules 4:564
24. Overberger CG, Chang JY, Gunn VE (1989) J Polym Sci Chem Ed 27:99
25. Overberger CG, Chang JY (1989) J Polym Sci Chem Ed 27:4013
26. Seita T, Yamauchi K, Kinoshita M, Imoto M (1972) Macromol Chem 154:263
27. Seita T, Yamauchi K, Kinoshita M, Imoto M (1973) Macromol Chem 164:15
28. Han MJ, Kim KH, Cho TJ, Choi KB (1990) J Polym Sci Chem Ed 28:2719
29. Han MJ, Choi KB, Chae JP, Hahn BS (1990) J Bioactive Compatible Polym 5:80
30. Han MJ, Choi KB, Chae JP, Hahn BS (1990) Bull Korean Chem Soc 11:154
31. Morris CJOR (1976) Separation methods in biochemistry. Wiley, New York
32. Tinoco I Jr (1961) J Am Chem Soc 83:5047
33. Rhodes W (1961) J Am Chem Soc 83:3609
34. Bush CA (174) In: Ts'o POP (ed) Basic principles in nucleic acid chemistry. Academic, New York, vol II, chap 2
35. Browne DT, Eisinger J, Leonard NJ (1968) J Am Chem Soc 90:7302
36. Inaki Y, Renge T, Kondo K, Takemoto K (1975) Makromol Chem 176:2683
37. MacDonald JR, Echols WE, Price TR, Fox RB (1972) J Chem Phys 57:1746
38. Iyata T, Ochiai H, Ueda K, Imamura A (1993) Macromolecules 26:6021
39. Riley M, Mailing B, Charmberin MJ (1966) J Mol Biol 20:359
40. Stevens CL, Felsenfeld G (1964) Biopolymers 2:293
41. Zmudzka B, Bollum FJ, Shugar D (1969) J Mol Biol 46:169
42. Blake RD, Massoulie J, Fresco JR (1967) J Mol Biol 30:291
43. Sequaris J-M, Koglin E, Valenta P, Nurnberg HW (1981) Ber Bunsen-Ges Phys Chem 85:512
44. Cotton TM (1988) In: Clark RJH, Hester RE (eds) Spectroscopy of surfaces. Wiley, New York, p 91
45. Brabec V, Niki K (1985) Biophys Chem 23:63
46. Kaye H (1970) J Am Chem Soc 92:5777
47. Akashi M, Okamoto T, Inaki Y, Takemoto K (1979) J Polym Sci Chem Ed 17:905
48. Takemoto K, Tahara H, Yamada A, Inaki Y, Ueda N (1973) Makromol Chem 169:327
49. Doel MT, Jones AS, Walker RT (1974) Tetrahedron 30:2755
50. Buttrey JD, Jones AS, Walker RT (1975) Tetrahedron 31:73
51. Draminski M, Pitha J (1978) Makromol Chem 179:2195
52. Ishikawa T, Inaki Y, Takemoto K (1978) Polym Bull 1:215
53. Takemoto K, Yahara H, Yamada A, Inaki Y, Ueda N (1973) Makromolek Chem 169:327

54. Alberts B, Bray D, Lewis J, Raff M, Roberts K, Watson JD (1994) Molecular biology of the cell, 3rd edn. Garland, New York
55. Uhlmann E, Peyman A (1990) Chem Rev 90:543
56. Nielsen PE, Egholm M, Berg RH, Buchardt O (1991) Science 254:1497
57. Egholm M, Buchardt O, Nielsen PE, Berg RH (1992) J Am Chem Soc 114:1895
58. Egholm M, Buchardt O, Nielsen PE, Berg RH (1992) J Am Chem Soc 114:9677
59. Egholm M, Buchardt O, Christensen L, Behrens C, Freier SM, Driver D, Berg RH, Kim SK, Nordén B, Nielsen PE (1993) Nature 365:566
60. Kim SK, Nielsen PE, Egholm M, Buchardt O, Berg RH, Nordén B (1993) J Am Chem Soc 115:6477
61. Wittung P, Nielsen PE, Buchardt O, Egholm M, Nordén B (1994) Nature 368:561
62. Wittung P, Eriksson M, Lyng R, Nielsen PE, Nordén B (1995) J Am Chem Soc 117:10,167
63. Wittung P, Nielsen PE, Nordén B (1997) J Am Chem Soc 119:3189
64. Cantin M, Schütz R, Leumann CJ (1997) Tetrahedron Lett 38:4211
65. Kosynkina L, Wang W, Liang YC (1994) Tetrahedron Lett 35:5173
66. Goodnow RA Jr, Richou A, Tam S (1997) Tetrahedron Lett 38:3195
67. Goodnow RA Jr, Tam S, Pruess DL, McComas WW (1997) Tetrahedron Lett 38:3199
68. Howarth NM, Wakelin LPG (1997) J Org Chem 62:5441
69. Kita Y, Uno T (1981) J Polym Sci Polym Chem Ed 19:477
70. Suda Y, Inaki Y, Takemoto K (1984) J Polym Sci Polym Chem Ed 22:623
71. Kunkel TA (1984) Proc Natl Acad Sci USA 81:1494
72. Vousden KH, Bos JL, Marsheall CJ, Phillips DH (1986) Proc Natl Acad Sci USA 83:1222
73. Lindahl T, Nyberg B (1972) Biochemistry 11:3610
74. Lindahl T, Andersson A (1972) Biochemistry 11:3618
75. Lindahl T (1982) Annu Rev Biochem 51:61
76. Drake JW, Baltz RH (1976) Annu Rev Biochem 45:11
77. Shaaper RN, Leob LA (1981) Proc Natl Acad Sci USA 78:1773
78. Han MJ, Cho TJ, Lee GH, Yoo KS, Park YD, Chang JY (1999) J Polym Sci Chem Ed 37:3361
79. Garrett ER, Seydel JK, Sharpen AJ (1966) J Org Chem 31:2199
80. Shapiro R, Danzig M (1972) Biochemistry 11:23
81. Shapiro R, Danzig M (1973) Biochim Biophys Acta 319:5
82. Cech TR (1990) Angew Chem Int Ed Engl 29:759
83. Altman S (1990) Angew Chem Int Ed Engl 29:749
84. Uhlenbeck OC (1987) Nature 328:596
85. Cech TR (1986) Sci Amer 255:76
86. Zuag AJ, Cech TR (1986) Science 231:470
87. McSwiggen JA, Cech TR (1989) Science 244:679
88. Cech TR (1987) Science 236:1532
89. Brehm SL, Cech TR (1983) Biochemistry 22:2390
90. Zaug AJ, Been MD, Cech TR (1986) Nature 324:429
91. Bass BL, Cech TR (1984) Nature 308:820
92. Guerrier-Takada C, Haydock K, Allen L, Altman S (1986) Biochemistry 25:1509
93. Guerrier-Takada C, Gardiner K, Marsh T, Pace N, Altman S (1983) Cell 35:849
94. Baer M, Altman S (1985) Science 228:999
95. Sribivasan A, Reddy EP, Dunn CY, Aaronson SA (1984) Science 223:285
96. Herschlag D, Cech TR (1990) Nature 344:405
97. Robertson DL, Joyce GF (1990) Nature 344:467
98. Han MJ, Macromolecules (in press)
99. Han MJ, Yoo KS, Cho TJ, Chang JY, Cha YJ, Nam SH (1997) Chem Comm 163
100. Han MJ, Yoo KS, Kim KH, Lee GH, Chang JY (1997) Macromolecules 30:5408

101. Nagae S, Miyamoto T, Inaki Y, Takemoto K (1989) Polym J 21:19
102. Nagae S, Suda Y, Inaki Y, Takemoto K (1989) J Polym Sci Polym Chem Ed 27:2593
103. Inaki Y, Ebisutani K, Takemoto K (1986) J Polym Sci Polym Chem Ed 24:3249
104. Inaki Y, Takemoto K (1988) J Macromol Sci Chem 25:757
105. Moghaddam MJ, Hozumi S, Inaki Y, Takemoto K (1989) Polym J 21:203
106. Moghaddam MJ, Hozumi S, Inaki Y, Takemoto K (1988) J Polym Sci Polym Chem 26:3297
107. Haines DS, Strauss KI, Gillespie DH (1991) J Cell Biochem 46:9
108. Torrence PF, DeClercq E (1990) Pharmacol Ther 2:1

Editor: Prof. K.-S. Lee
Received: January 2000

Structure and Applications of Poly(vinyl alcohol) Hydrogels Produced by Conventional Crosslinking or by Freezing/Thawing Methods

Christie M. Hassan, Nikolaos A. Peppas

Polymer Science and Engineering Laboratories, School of Chemical Engineering, Purdue University, West Lafayette, IN 47907–1283, USA

Poly(vinyl alcohol) (PVA) is a polymer of great interest because of its many desirable characteristics specifically for various pharmaceutical and biomedical applications. The crystalline nature of PVA has been of specific interest particularly for physically crosslinked hydrogels prepared by repeated cycles of freezing and thawing. This review includes details on the structure and properties of PVA, the synthesis of its hydrogels, the crystallization of PVA, as well as its applications. An analysis of previous work in the development of freezing and thawing processes is presented focusing on the implications of such materials for a variety of applications. PVA blends that have been developed with enhanced properties for specific applications will also be discussed briefly. Finally, the future directions involving the further development of freeze/thawed PVA hydrogels are addressed.

Keywords. Poly(vinyl alcohol), Hydrogels, Freezing/thawing, Crystallinity, Biomedical applications, Crosslinking, Crystallites

1	Structure and Properties of Poly(vinyl alcohol) (PVA)	38
2	Synthesis and Properties of PVA Hydrogels	39
3	Crystallization of PVA Hydrogels	41
4	Hydrogels by Freezing and Thawing of Aqueous PVA Solutions	45
5	PVA Blends by Conventional and Freezing/Thawing Techniques	55
6	Biomedical and Pharmaceutical Applications of PVA	57
7	Summary	61
References		62

1
Structure and Properties of Poly(vinyl alcohol) (PVA)

Poly(vinyl alcohol) (PVA) has a relatively simple chemical structure with a pendant hydroxyl group. The monomer, vinyl alcohol, does not exist in a stable form rearranging to its tautomer, acetaldehyde. Therefore, PVA is produced by the polymerization of vinyl acetate to poly(vinyl acetate) (PVAc), followed by hydrolysis of PVAc to PVA. The hydrolysis reaction does not go to completion resulting in polymers with a certain degree of hydrolysis that depends on the extent of reaction. In essence, PVA is always a copolymer of PVA and PVAc. Commercial PVA grades are available with high degrees of hydrolysis (above 98.5%). The degree of hydrolysis, or the content of acetate groups in the polymer, has an overall effect on its chemical properties, solubility, and the crystallizability of PVA [1].

The degrees of hydrolysis and polymerization affect the solubility of PVA in water [2]. It has been shown that PVA grades with high degrees of hydrolysis have low solubility in water. Figure 1 shows the solubility of a PVA sample with a number average molecular weight of $\overline{M}_n$=77,000 as a function of the degree of hydrolysis at dissolution temperatures of 20 and 40 °C. Residual hydrophobic acetate groups weaken the intra- and intermolecular hydrogen bonding of adjoining hydroxyl groups. The temperature must be raised well above 70 °C for dissolution to occur. The presence of acetate groups also affects the ability of PVA to crystallize upon heat treatment. PVA grades containing high degrees of hydrolysis are more difficult to crystallize.

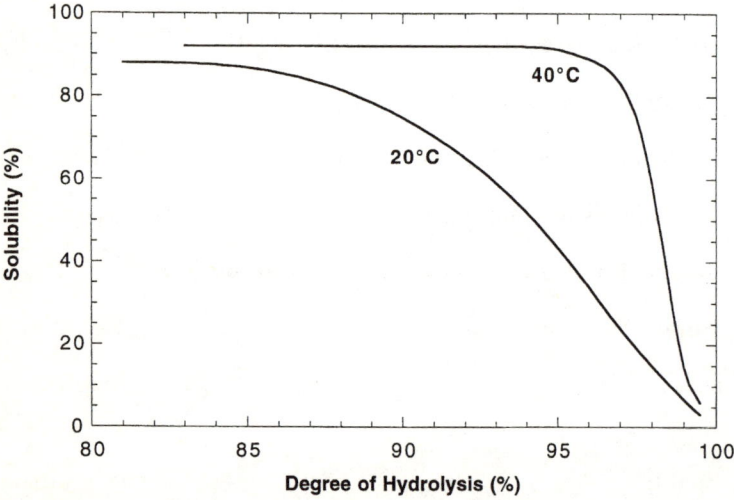

Fig. 1. Solubility as a function of degree of hydrolysis at dissolution temperatures of 20 and 40 °C

PVA is produced by free radical polymerization and subsequent hydrolysis, resulting in a fairly wide molecular weight distribution. A polydispersity index of 2 to 2.5 is common for most commercial grades. However, polydispersity indices of 5 are not uncommon. The molecular weight distribution is an important characteristic of PVA because it affects many of its properties including crystallizability, adhesion, mechanical strength, and diffusivity.

2
Synthesis and Properties of PVA Hydrogels

PVA must be crosslinked in order to be useful for a wide variety of applications, specifically in the areas of medicine and pharmaceutical sciences. A hydrogel can be described as a hydrophilic, crosslinked polymer (network) which swells when placed in water or biological fluids [3]. However, it remains insoluble in solution due to the presence of crosslinks.

PVA can be crosslinked through the use of difunctional crosslinking agents [4]. Some of the common crosslinking agents that have been used for PVA hydrogel preparation include: glutaraldehyde, acetaldehyde, formaldehyde, and other monoaldehydes. When these crosslinking agents are used in the presence of sulfuric acid, acetic acid, or methanol, acetal bridges form between the pendant hydroxyl groups of the PVA chains. As with any crosslinking agent, however, residual amounts are present in the ensuing PVA gel. It becomes extremely undesirable to perform time-consuming extraction procedures in order to remove this residue. If the residue is not removed, the gel will not be acceptable for biomedical or pharmaceutical applications. If the gel were to be used in biomedical applications, the release of this toxic residue would have obvious undesirable effects. For pharmaceutical applications, especially when PVA is used as a carrier in drug delivery, the toxic agent could alter the biological activity or degrade the biologically active agent being released. There are also other toxic residual components associated with chemical crosslinking such as initiators, chain transfer agents, and stabilizers.

Other methods of chemical crosslinking include the use of electron beam or γ-irradiation. These methods have advantages over the use of chemical crosslinking agents as they do not leave behind toxic, elutable agents. Danno [5] discussed the quantitative and qualitative effects of irradiation by γ-rays (from ^{60}Co sources) to produce hydrogels from aqueous PVA solutions. He found that the minimum gelation dose depended on the degree of polymerization and the concentration of the polymer in solution. Further studies by Saito [6] and Dieu and Desreux [7] showed that when no impurities are present, the intrinsic viscosity and the viscosity-average molecular weight of irradiated aqueous solutions of PVA increase with radiation dose.

Sakurada and Mori [8] and Peppas and Merrill [9] investigated the effects of γ-irradiation by ^{60}Co on the physical properties of PVA fibers, hydrogels, and films irradiated in water. The effect of the irradiation dose on the molecular weight between crosslinks is more closely represented in Fig. 2 for aqu-

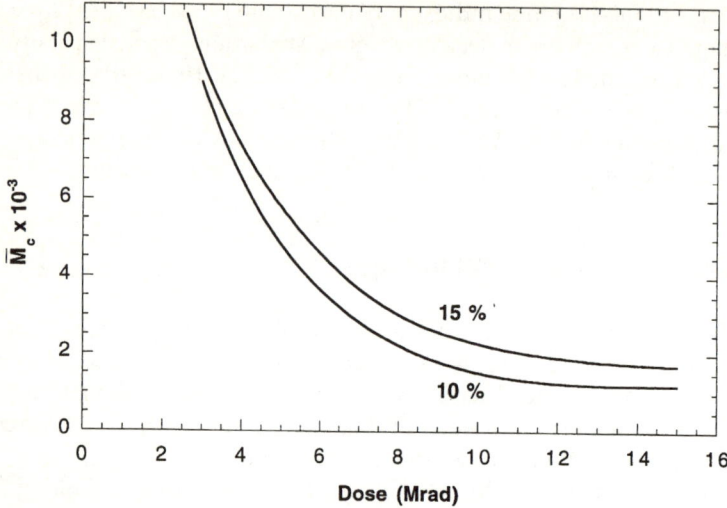

Fig. 2. Effect of concentration and irradiation dose on the molecular weight between crosslinks $\overline{M}_c$

eous PVA solutions of 10 and 15 wt %. Significant differences in swollen films were observed between irradiation under vacuum and in air. However, when the concentration of the aqueous solution was significantly high, little difference was observed. One problem observed in this technique of crosslinking was bubble formation. This problem was also encountered by other researchers [5, 10] with little success in solving the problem. A proposed solution was to pre-cool the aqueous solution for a certain amount of time to yield homogeneous hydrogels. This, however, is not a desirable procedure to carry out with PVA.

The mechanical strength of crosslinked, partially crystalline PVA hydrogels has already been examined [11]. In particular, efforts were made to enhance the mechanical strength of crosslinked PVA networks that were prepared initially by electron beam irradiation of aqueous PVA solutions. Peppas and Merrill [9, 12] examined the phenomenon of partial crystallization by a process of dehydration and annealing. Resulting materials contained crystallites in addition to crosslinks. The crystalline regions essentially served as additional crosslinks to redistribute external stresses. The degree of crystallinity of samples annealed at 120 °C for 30 min is shown as a function of the crosslinking density in Fig. 3.

A third mechanism of hydrogel preparation involves "physical" crosslinking due to crystallite formation [13]. This method addresses toxicity issues because it does not require the presence of a crosslinking agent. Such physically crosslinked materials also exhibit higher mechanical strength than PVA gels crosslinked by chemical or irradiative techniques because the mechanical load can be distributed along the crystallites of the three-dimensional struc-

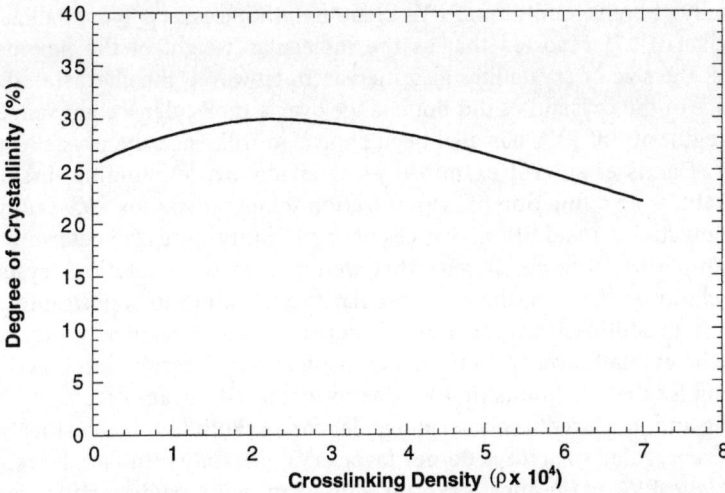

Fig. 3. Degree of crystallinity as a function of crosslinking density of PVA gels swollen to equilibrium in water at 37 °C

ture. Aqueous PVA solutions have the unusual characteristic of crystallite formation upon repeated freezing and thawing cycles. The number and stability of these crystallites are increased as the number of freezing/thawing cycles is increased. Some characteristics of these "physically" crosslinked PVA gels include a high degree of swelling in water, a rubbery and elastic nature, and high mechanical strength. In addition, the properties of the gel may depend on the molecular weight of the polymer, the concentration of the aqueous PVA solution, the temperature and time of freezing and thawing, and the number of freezing/thawing cycles. The development of this technique and its ensuing materials will be discussed in more detail in Sect. 4.

3
Crystallization of PVA Hydrogels

The crystalline structure of PVA has been discussed in detail by Bunn [14]. On a molecular level, the crystallites of PVA can be described as a layered structure [15, 16]. A double layer of molecules is held together by hydroxyl bonds while weaker van der Waals forces operate between the double layers. A folded chain structure of PVA chains leads to small ordered regions (crystallites) scattered in an unordered, amorphous polymer matrix. Values representative of the crystallinity and thermal properties of PVA have been reported [4]. The crystalline melting range of PVA is between 220 and 240 °C. The glass transition temperature of dry PVA films has been reported at 85 °C. In the presence of water (and other solvents), the glass transition temperature decreases significantly [2].

Usually, there is a minimum PVA chain length necessary to crystallize PVA. Mandelkern [17] reported that as the molecular weight of the polymer increased, the size of crystallites also increased. However, the character and appearance of the crystallites did not change over a molecular weight range. The stereoregularity of PVA has also been shown to influence the crystallizability of PVA. Harris et al. [18] examined these effects by determining dissolution temperatures as a function of crystallization temperatures for PVA samples of varying tacticity. In addition, degrees of crystallinity were also determined for PVA samples of varying tacticity that were prepared by solution-crystallization techniques. Overall, the stereoregularity was found to significantly affect the crystallizability. However, PVA of increasing stereoregularity did not increase the crystallizability. In particular, isotactic PVA resulted in less crystalline samples than syndiotactic PVA due to increased intramolecular hydrogen bonding and, thus, reduced intermolecular forces. Fujii [19] has further shown that stereoregular structures do not favor crystallizability. In fact, he reported that atactic PVA is the most crystallizable form, with syndiotactic somewhat less crystallizable, with the isotactic form showing poor crystallinity.

Different methods have been investigated for the crystallization of PVA. Most of these methods involve heat treatment to introduce crystallites. The degree of crystallinity, as well as the size of crystallites, depends on the drying conditions. One method involves slow-drying rate dehydration at room temperature under different relative humidities [20–22]. In this process, the amount of polymer in the system is increased and crystallization actually begins late in the process. The effect of drying time on the degree of crystallinity is shown in Fig. 4 where dehydration of the samples was performed with

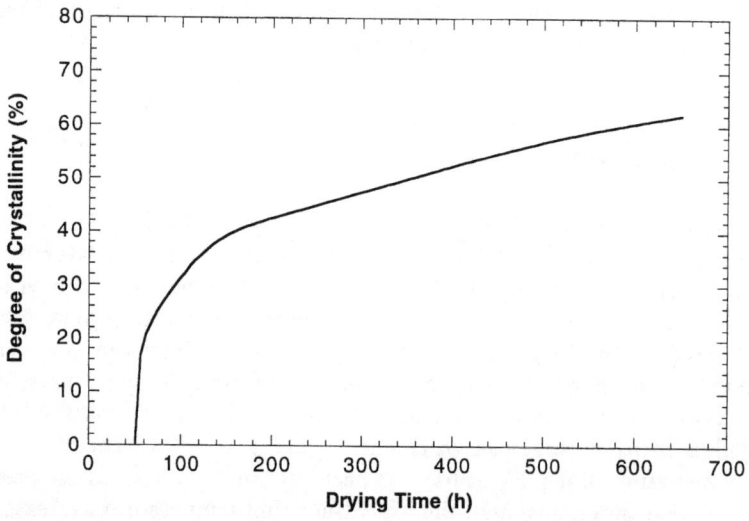

Fig. 4. Degree of crystallinity of dehydrated PVA gels as a function of drying time

$Ca(NO_3)_2$ as the drying agent. Fast-drying rate processes, such as annealing, usually result in the rapid formation of crystallites. One such method involves annealing PVA films at a constant temperature that is between the glass transition temperature and the melting point of the polymer [23]. Under these conditions, macromolecular chains have enough mobility to align themselves and fold to form crystallites. Gels produced by such techniques involving these annealing conditions have been further characterized recently in terms of their structure and dissolution behavior in water [24].

Yamaura et al. [25–27] examined PVA gels that had been prepared in the presence of various solvents including dimethylformamide, ethylene glycol, and phenol. The thermal and mechanical properties of these gels and dried gel films were studied. Additionally, gel spinning was implemented with solutions of PVA, phenol, and water to produce filaments with very high dynamic moduli (up to 45 GPa).

It has also been shown that crystallization takes place in PVA gels due to aging [28]. A decrease in the swelling ratio in water is indicative of this phenomenon which has been confirmed by X-ray analysis. Tanigami et al. [29] have more closely characterized aged PVA gels. Such gels were prepared in a mixed solvent of dimethyl sulfoxide (DMSO) and water. PVA films were aged in water for several months at 10 °C, then dried, and drawn at 200 °C. The aging process significantly affected the tensile strength and modulus of the drawn films. Specifically, the low temperature of 10 °C was necessary to allow for crystallites at low melting temperature to grow within the aged gel. Additional work by Tanigami et al. [30] more closely examined the swelling of PVA gels that had been dried and annealed. The swelling agent was a mixed solvent of DMSO and water. They characterized the relaxation and swelling rates of such gels in relation to the solvent composition and degree of crystallinity. Anomalous swelling was attributed to the low relaxation rate of PVA that compared with the diffusion coefficient of the penetrant. Another key observation was that the degree of crystallinity played an important role in controlling the relaxation rate. Further work by Tanigami et al. [31] also characterized the aging of PVA gels prepared by freezing and thawing techniques and will be discussed in Sect. 4.

Semicrystalline PVA is characterized by its degree of crystallinity which is defined as the ratio of the volume of PVA crystallites to the total volume of PVA. Density measurements, calorimetric methods, spectroscopic methods, X-ray analysis, and other methods such as NMR spectroscopy have been used to determine degrees of crystallinity of PVA hydrogels [32]. Using such methods, degrees of crystallinity were determined for PVA samples ($\overline{M}_c$=4770) that were prepared under varying annealing conditions (Table 1).

Density measurements can be used for PVA hydrogels without the presence of voids or other imperfections. The calculation of the degree of crystallinity uses the principle of additivity of volumes and the densities of 100% crystalline and amorphous PVA [15]. These values have been reported in the literature by Sakurada et al. [34] as ρ_c=1.345 g/cm^3 for 100% crystalline PVA and

Table 1. Crystallinity as a function of preparation conditions as determined from various techniques

Annealing conditions		Degree of crystallinity (%)		
Temp (°C)	Time (min)	From density	From DSC	From IR spectra
90	30	30.0	35.6	31.1
105	60	45.0	46.8	44.5
120	30	50.8	50.1	52.3
120	90	64.7	71.1	62.7

$\rho_a = 1.269$ g/cm^3 for 100% amorphous PVA. The degree of crystallinity, X, of PVA on a dry basis is:

$$\frac{1}{\rho_g} = \frac{X}{\rho_c} + \frac{1-X}{\rho_a} \qquad (1)$$

where ρ_g is the density of the sample. This can be modified for PVA hydrogels where the additivity of the volumes of PVA and water is assumed:

$$\frac{1}{\rho_h} = \frac{1-W_{PVA}}{\rho_{H_2O}} + \frac{W_{PVA}}{\rho_g} \qquad (2)$$

Here, ρ_h is the density of the hydrogel, W_{PVA} is the weight percentage of PVA in the sample, and ρ_{H_2O} is the density of water.

Calorimetric methods can also be used to determine the crystallinity of PVA films. They include differential thermal analysis, differential scanning calorimetry (DSC), and thermogravimetric analysis. Typically, the heat of crystallization of a PVA sample, ΔH, is compared to the heat of crystallization of 100% crystalline PVA, ΔH_c. The degree of crystallinity is expressed as:

$$X = \frac{\Delta H}{\Delta H_c} \qquad (3)$$

The melting of PVA occurs over a range of temperatures because the crystallites range in size. Therefore, calorimetric methods can also be used to calculate crystallite size distributions.

Of the spectroscopic methods, infrared (IR) spectroscopy is the most widely used. The intensity of the peak at 1141 cm^{-1} in the IR spectrum of PVA depends on the degree of crystallinity. It has been proposed that the band arises from a C–C stretching mode and increases with an increase in the degree of crystallinity [34]. A method for determining degrees of crystallinity from ATR-FTIR spectra has been further discussed by Peppas [35]. The height of the peak at 1141 cm^{-1} is analyzed in terms of the height of the peak at 1425 cm^{-1} which remains constant.

X-ray analysis has also been used for films and fibers of PVA. X-ray diffraction methods are used to determine the percentage of crystalline PVA by comparing the scattering of a sample to that of a completely amorphous one. However, in the case of swollen PVA films, results are difficult to analyze due to the existence of the pattern of water. Sakurada et al. [33] analyzed the crystallinity of swollen PVA films, showing that swelling does not change ("melt out") the crystalline regions of PVA.

4
Hydrogels by Freezing and Thawing of Aqueous PVA Solutions

To avoid crosslinking processes which potentially lead to the release of toxic agents, a physical method of gelation and solidification of some polymers, particularly PVA, has been developed. This freezing and thawing process of PVA gel preparation has been summarized recently by various researchers [36, 37]. In particular, the impact of various preparation conditions on the overall properties of such materials has been addressed. Here, a thorough analysis of previous work in this area will be presented. In particular, the implications of such materials prepared by freezing and thawing processes for various applications will be described.

The preparation of pure PVA hydrogels using freezing and thawing techniques was first reported by Peppas [38] in 1975. In this work, aqueous solutions of between 2.5 and 15 wt% PVA were frozen at $-20\,°C$ and thawed back to room temperature resulting in the formation of crystallites. The crystallites were characterized using measurements of the change in turbidity of the PVA samples. The formation of crystallites in the samples was found to be related to the concentration of PVA in solution, the freezing time, and the thawing time. The transmittance of visible light through samples prepared with vary-

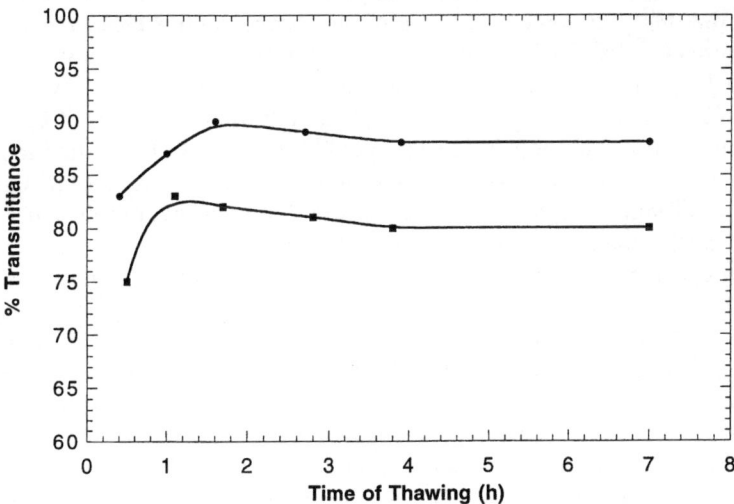

Fig. 5. Transmittance of visible light as a function of thawing time for samples prepared from 10 (●) and 15% (■) aqueous PVA solutions

ing aqueous PVA concentrations is shown as a function of thawing time in Fig. 5. Overall, the crystallinity was found to increase with increasing freezing time. During the thawing process, the size of the crystallites initially increased and then decreased. This was attributed to the breakdown of the crystalline structure. The degree of crystallinity was found to increase with increasing PVA solution concentration.

Since this pioneering discovery, much research has been conducted on the freezing and thawing process as well as on the characterization of such gels produced with these techniques. The research of Nambu [39] introduced the potential use of pure PVA gels produced by freezing and thawing processes for biomedical applications. His freeze-drying technique consisted of cooling a PVA solution to below $-3\,°C$ followed by the vacuum evaporation of water. Further work by Nambu and collaborators [40] examined the mechanical properties of gels prepared with this freeze-drying process. Simple extension experiments were performed to analyze the force-extension relationship at large deformations. It was determined that an increase in the evacuation (evaporation of water) time resulted in an increase in the tensile force. This was attributed to the strengthening of the gel by decreasing the number of imperfections which may not have been attached to the network structure. More recently, Nambu and collaborators [41] have also examined PVA gels prepared by freezing and thawing techniques for use as a phantom in magnetic resonance imaging due to its close resemblance to human soft tissue.

Yokoyama et al. [42] further examined gels that had been prepared by Nambu's method of repeated freezing and thawing cycles. Using X-ray diffraction, scanning electron microscopy (SEM), light-optical microscopy, and tension experiments, the structure was described as one consisting of three phases: a water phase of low PVA concentration, an amorphous phase, and a crystalline phase that restricts some of the motion of the amorphous PVA chains.

Watase and Nishinari [43] further characterized freeze/thawed gels through the analysis of DSC endotherm peaks and X-ray diffraction. They also performed an extensive analysis concerning the effect of the degree of hydrolysis on various properties of PVA gels prepared by freezing and thawing [44]. They reported that the presence of bulky acetate groups can inhibit the formation of a gel. Additionally, slight differences in the degree of hydrolysis can significantly impact the thermal behavior of the gels.

In preparing such PVA gels by freezing and thawing techniques, the addition of solvents has also been investigated. Hyon and Ikada [45] prepared porous and transparent hydrated gels from a PVA solution in a mixed solvent of water and water-miscible organic solvents. The concentration of PVA in solution was between 2 and 50 wt %. The organic solvents examined included dimethyl sulfoxide, glycerine, ethylene glycol, propylene glycol, and ethyl alcohol. The method consisted of cooling the solution to below $0\,°C$ for the crystallization of PVA followed by the subsequent exchange of the organic solvent in the gel with water. This process resulted in the formation of a hydrated gel of PVA with high tensile strength, high water content, and high light transmittance. High light transmittance was an important property of the material, especially when considering the material for contact lens applications. Explanations were offered concerning the mechanism for gel formation as well as increased light transmittance. As the temperature of the homogeneous solution was lowered, there was a restriction in the molecular motion. The intermolecular interaction of PVA, likely due to hydrogen bonding, was promoted to yield small crystalline nuclei. Crystallization could proceed further as the solution remained at low temperatures for longer times. The crystallites served as crosslinks to hold the three-dimensional structure together. The addition of the organic solvent served to prevent the PVA solution from freezing even below 0 °C. This allowed for PVA crystallization to proceed without significant volume expansion. Therefore, the ensuing gel contained pores of less than 3 μm, resulting in a transparent gel. Additional work by Hyon et al. [46] further examined the preparation of such PVA gels due to crystallization at low temperatures.

Other contributions have been made concerning the characterization of PVA gels prepared by freezing and thawing techniques in the presence of other solvents. In particular, Berghmans and Stoks [47] described a two-step mechanism for the gelation of PVA gels in the presence of ethylene glycol. The first step involves a liquid-liquid phase separation that occurs as the gel turns opaque. Then during the second step, crystallization occurs in the more concentrated phase.

Lozinsky and collaborators [48, 49] characterized various aspects of cryogels, or gels prepared by freezing and thawing processes. Initial work showed that upon exposing aqueous PVA to one cycle of freezing and thawing, no destruction or covalent crosslinking of macromolecules of PVA was noted. An extensive investigation of freeze/thawed PVA gels involved the introduction of a filler that could model microbial cells. The filler that was used consisted of spherical particles of a crosslinked dextran gel. Such gels were characterized in terms of their rheological and thermal properties. Additional work in this group [50, 51] examined various properties of freeze/thawed gels prepared in the presence of other components. The influence of both polyols and low molecular weight electrolytes was investigated. The addition of triethylene glycols and higher oligomers of ethylene glycol was found to increase the strength and thermal stability of the gels. Freezing and thawing in the presence of electrolytes was also found to impact the gelation process and subsequent swelling characteristics.

PVA hydrogels prepared by freezing and thawing techniques were further characterized by Urushizaki et al. [52, 53], and it was determined that many properties of these gels were closely related. They found a good correlation between tack and viscoelasticity which was useful for predicting the properties of the material for pressure-sensitive adhesives in transdermal drug-delivery systems. Additionally, the amount of unincorporated PVA, the density of the PVA gel, viscoelasticity, and swelling behavior were also closely related. The rate of swelling of the gels was found to increase linearly with the square root of time immersed in water. In addition, the water uptake upon swelling in water increased with increasing temperature. In studying the mechanical properties, the viscoelasticity showed little change between 15 and 50 °C but changed significantly above 50 °C as there was an irreversible physical change from gel to sol.

Further research by Ohkura et al. [54] focused on the use of dimethyl sulfoxide/water solutions. They reported interesting features of PVA gels that were formed from solutions of mixtures of DMSO and water. Gels that were prepared by freezing below 0 °C were transparent and exhibited high elasticity. Higher gelation rates were observed when compared with aqueous solutions of only PVA in water. Specifically, the properties were dependent on the ratio of dimethyl sulfoxide to water. It was also observed that gelation from the mixture occurred without phase separation below −20 °C. However, above this temperature, phase separation plays an important role for the gelation process. This research emphasized that the structure of the gels was not well understood from microscopic viewpoints leading to a sparked interest in studying the gels by wide- and small-angle neutron scattering. Additional work by this group [55] determined the gelation rate as a function of polymer concentration, quenching temperature, and degree of polymerization.

In this continued work [56–58], it was confirmed by wide-angle neutron scattering that crosslinking, or points of junction, in the gels were indeed crystallites. Furthermore, small-angle neutron scattering experiments showed that the surface of the crystallites had a clear boundary. The average size of the crystallites was approximately 70 Å and the average distance between crystallites was approximately 150 to 200 Å

The thermal properties of gels prepared by freezing and thawing techniques were further examined by Hatakeyama et al. [59]. They classified the water contained within the gels as non-freezing, restrained, and free water. In addition, the work showed that the rate of freezing affected the size of ice crystals formed and, therefore, the number of crosslinks formed.

Cha et al. [60] further examined the microstructure of PVA gels with differential scanning calorimetry. They also classified the water in PVA hydrogels as free, intermediate, or bound water. PVA gels were prepared by three different techniques: chemical crosslinking with glutaraldehyde, annealing of dried films, and crystallization at low temperatures. Gels prepared by low temperature crystallization contained less bound water with the concentration increasing as the degree of polymerization of PVA was increased. This was likely attributed to the fact that the concentration of free PVA chains, which did not participate in the crystallite formation process, increased with the increasing degree of polymerization. It was determined that low temperature crystallization resulted in larger free space between crystallites and larger size of crystallites than annealing.

Significant changes in the crystalline structure of freeze/thawed gels over time was examined by Murase et al. [61]. They studied the syneresis, or volumetric shrinkage (i.e. solvent exclusion), of gels prepared from PVA solutions in a mixed solvent of DMSO and water. They attributed increased crystallinity over time to be due to the interaction between DMSO and water which isolated PVA hydroxyl groups. Further work by this group [31] studied the structure of aged PVA gels and the associated syneresis. The gels were prepared from PVA solutions in a mixed solvent of DMSO and water by freezing at $-34\,°C$ for one day. The gels were aged for up to 500 days at $30\,°C$ to allow for the growth of crystallites. The aged gel was found to have an increased modulus as well as solvent exclusion. This was again attributed to a densification of the structure caused by phase separation. The crystallization produced three components which were detected in the form of a multi-peak endotherm using differential scanning calorimetry.

Trieu and Qutubuddin [62] further examined the structure of freeze/thawed PVA gels prepared from aqueous DMSO solutions. In particular, they characterized the gels using freeze-etching and critical point drying SEM techniques. A higher porosity was observed at the surface than within the bulk of the gel. Tazaki et al. [63] also examined the porous nature of freeze/thawed PVA gels through SEM techniques. They confirmed the presence of fairly large pores within the structure. They further examined such gels for the separation of a water/ethanol mixture by pervaporation.

The porous structure of freeze/thawed PVA gels has been further characterized by Suzuki et al. [64, 65]. Gels containing poly(acrylic acid) and poly(allyl amine) have porous structures that result in the fast exchange of solvents. The relationship between the mechanical properties and porosity was more closely examined. A substantial increase in the modulus was observed with increasing cycles of freezing and thawing for up to ten cycles. Above ten cycles, the modulus still increased but not as significantly. Additional work focused on cryo-SEM techniques and X-ray analysis to characterize the porous structure of amphoteric PVA gels. The influence of the cooling rate and the number of freezing and thawing cycles was further examined.

There have also been applications of such freeze/thawed materials in the biomedical field. Bao and Higham [66] reported the use of such materials as intervertabrate disc nuclei. Specifically, the hydrogels prepared had a water content of 30% or higher and a compressive strength of 4 MPa or higher. They discussed the combination of various pieces of the hydrogel to obtain an overall conformation of the natural disc nucleus.

Recently, the use of freeze/thawed PVA hydrogels for particular drug-release applications has been investigated. Takamura and collaborators [67–69] examined the use of PVA gels prepared by freezing/thawing processes for various applications including controlled drug release. Early work [67] first addressed the preparation of strong, freeze/thawed PVA gels. They reported that it was necessary for the grade of PVA to have a degree of polymerization of over 1000 and a concentration of PVA of over 6% to obtain adequate strength. In particular, the strength was found to increase with the first three cycles of freezing and thawing and then level off. They also characterized the drug release of PVA gels prepared from emulsions [68]. Release of both hydrophobic and hydrophilic drugs was found to be zero-order. Such a system was examined for use as a suppository. Additional work by this group [69] focused on studying the effects of additives on the release and physicochemical properties of the gel. The additives used were sodium alginate and Pluronic L-62. The physical strength of the gels was found to increase with increasing sodium alginate concentration. The addition of Pluronic L-62 had no effect on the overall strength. However, drug release from the gel was found to decrease with an increase in the content of sodium alginate or Pluronic L-52.

PVA gels prepared with a mixed solvent of water and DMSO were also characterized in terms of artificial articular cartilage applications [70]. The solutions were cooled to below room temperature to allow for gelation. The gels were then exposed to annealing at various temperatures. Such preparation techniques produced PVA gels with increased tensile strength and dynamic modulus which are important when considering the wear properties of the polymer.

Significant contributions to the development of PVA gels prepared by freezing and thawing techniques have been made in our own laboratory. Some of the parameters that have been investigated include the concentration of aqueous PVA, the molecular weight of PVA, the number of freezing and thawing cycles, the time of freezing, and the time thawing. Specifically, the gels investigated have been characterized in terms of swelling behavior, degree of crystallinity, the transport and release of drugs and proteins, mechanical strength, and adhesive and mucoadhesive characteristics. Some of the most recent characterizations have been described by Peppas [71] and addressed in terms of potential applications.

Stauffer and Peppas [13] investigated the preparation and properties of such materials. Aqueous solutions of 10 to 15 wt% PVA were frozen at –20 °C for between 1 and 24 h and then thawed at 23 °C for up to 24 h for five cycles. It was found that 15 wt% solutions produced strong thermoreversible gels with mechanical integrity. Specifically, gels that were frozen for 24 h for five cycles and thawed for any period of time were the strongest. Swelling experiments as a function of thawing time and freezing cycles indicated that denser structures were observed after five cycles (Fig. 6). This was indicative that the physical crosslinking, due to the presence of crystallites, increased with an increasing number of freeze/thaw cycles.

Peppas and Scott [72] investigated such PVA gels in terms of controlled release applications. PVA hydrogels were prepared by freezing aqueous solutions of 15 wt% PVA at –20 °C for 18 h and thawing at room temperature for 6 h for three, four, and five cycles. The samples with the highest PVA molecular weight and the highest number of freeze/thaw cycles led to the densest gel structure. These gels were also observed to be the strongest and most rigid. Swelling studies in conjunction with DSC experiments indicated that there was an initial decrease in the degree of crystallinity due to the melting out of smaller crystallites. This stage was followed by an increase in the crystallinity at long times which was attributed to additional crystallite formation due to

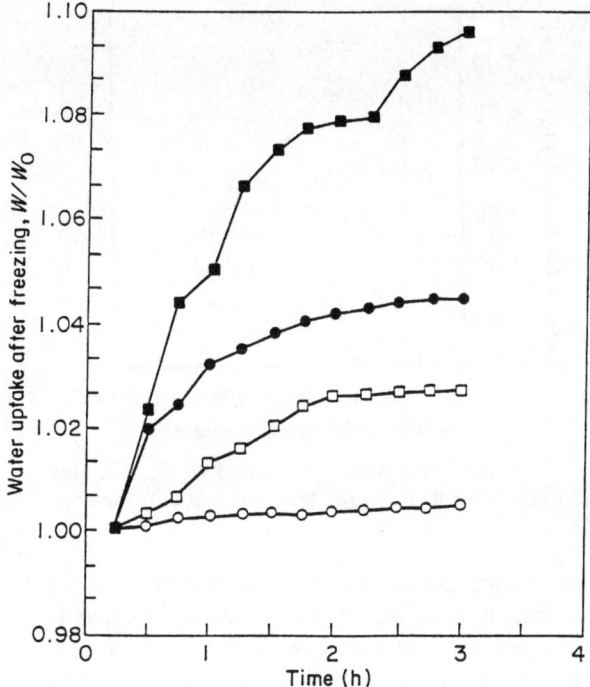

Fig. 6. Swelling in water at 23 °C for freeze/thawed gels prepared with two (■), three (●), four (□), and five (○) cycles of freezing and thawing

aging. The melting out of crystallites as well as secondary crystallization have proved to be important issues when considering the overall stability of this system. Release studies were also conducted with bovine serum albumin which was incorporated into the gel prior to freezing and thawing. The release profile was examined over a period of 19 days and determined to be Fickian. Therefore, bovine serum albumin release was controlled by diffusion with changes in crystallinity having little effect on the release mechanism.

Ficek and Peppas [73, 74] further examined controlled drug delivery from ultrapure PVA gels with the preparation of microparticles. PVA microparticles with diameter ranging from 150 to greater than 1400 μm were prepared by freezing and thawing processes. An aqueous solution of PVA was dispersed in corn oil with 1.25 wt% sodium lauryl sulfate as the surfactant. The suspended droplets of PVA solution were solidified upon exposure to several freezing/thawing cycles. Important parameters which were investigated in this work were the oil to PVA ratio, the amount of surfactant added, and the reagitation of particles after the oil had been partially frozen. The microparticles were investigated for bovine serum albumin release. High release rates were initially observed followed by a constant period of release over longer times (7 days).

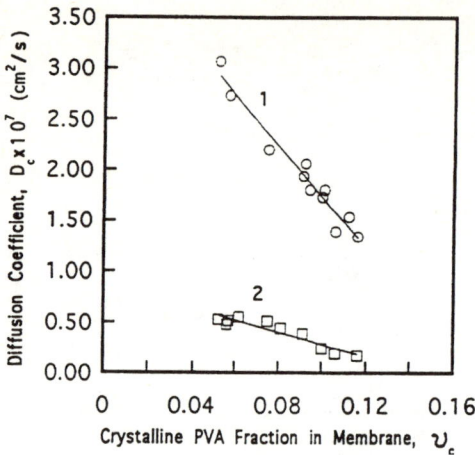

Fig. 7. Diffusion coefficients through freeze/thawed PVA gels as a function of the crystalline PVA volume fraction for theophylline (*curve 1*) and FITC-dextran (*curve 2*)

The diffusive characteristics of PVA gels prepared by freezing/thawing techniques were examined more closely by Hickey and Peppas [75]. Semicrystalline PVA membranes were prepared by freezing and thawing aqueous solutions of PVA for up to ten cycles under the conditions previously described. It was determined that the PVA crystalline fraction was a function of the number of cycles as well as the duration of each cycle. The volume-based crystalline fraction of PVA on a wet basis ranged from 0.052 to 0.116. The equilibrium volume swelling ratio also varied from 4.48 to 9.58 with decreasing degree of crystallinity. The focus of this work concerned diffusion studies. The permeation of theophylline and FITC-dextran was investigated with PVA membranes to determine size exclusion characteristics. It was determined that the diffusion coefficient could be approximated by a linear relation with the crystalline PVA volume fraction as shown in Fig. 7. It was also found that the transport of theophylline in semicrystalline PVA membranes produced by heat treatment was an order of magnitude higher than transport in the membranes under investigation. Overall, solute transport was a function of the crystalline fraction of PVA and the mesh size. Figure 8 shows in more detail the three-dimensional network structure of swollen, freeze/thawed PVA gels that was proposed in this work. The mesh size is shown between the crystalline regions that serve as the physical crosslinks. Overall, a size exclusion phenomenon was demonstrated by these permeation studies which was attributed to the crystallites that created the physical network.

The mucoadhesive characteristics of ultrapure PVA gels were examined in detail by Peppas and Mongia [76]. PVA hydrogels were prepared by exposing 15 to 20 wt% aqueous solutions of PVA to repeated cycles of freezing for 6 or 12 h at −20 °C and thawing for 2 h at 25 °C. The average degree of crystallinity for the 20 wt% solution was 19.3% on a dry basis. Adhesion studies showed

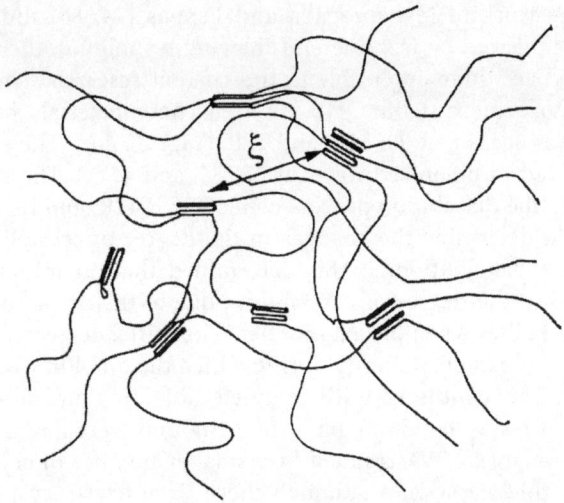

Fig. 8. Three-dimensional network structure of freeze/thawed PVA gels

that the work of fracture, or detachment, decreased with an increasing number of freezing/thawing cycles due to the increase in PVA degree of crystallinity. Maximum adhesion was achieved after two cycles. So, although increasing the number of freezing and thawing cycles likely contributed to the stability of the network, the adhesive characteristics decreased. Oxprenolol and theophylline were used in drug-release studies. Drug release was affected by the number of freezing/thawing cycles. Their results indicated that the mucoadhesive characteristics and drug release could be optimized for a certain mucoadhesive controlled release application by controlling freezing and thawing conditions.

Such bioadhesive materials were also examined for the controlled release of growth factors and other proteins [77–82]. Of particular interest was the investigation of mucoadhesive PVA gels for use in epidermal bioadhesive systems for the controlled release of epidermal growth factor or ketanserin to accelerate wound healing [78]. Such systems were examined not only for their drug-release properties but also for their strength and adhesive characteristics.

The most recent work on PVA gels prepared by freezing and thawing techniques [83–85] has focused on the effect of preparation conditions on the structure, morphology, and stability over long time periods. Parameters involved in the preparation of the PVA gels included the number of freezing and thawing cycles, the concentration of the aqueous solution, and the molecular weight of PVA. When considering the stability, several issues were addressed including the possible dissolution of PVA, melting out of crystalline regions, and additional crystallization during swelling at long times. In addition, structure and stability have been related to the use of such materials as carriers for drug delivery [85].

Although the work of Mallapragada and Peppas [24, 86] did not involve pure PVA gels prepared by freezing and thawing techniques, the analysis and results obtained are quite applicable to the current research. They examined the dissolution of semicrystalline PVA in water. These materials, however, were crystallized by annealing at 90, 110, and 120 °C for 30 min. The samples were partially dissolved in deionized water at 25, 35, and 45 °C. The samples were analyzed during the dissolution process using DSC, FTIR, and IR spectral analysis in order to determine the changes in the degree of crystallinity and lamellar thickness distribution. It was determined that an initial drastic decrease occurred in the degree of crystallinity due to the unfolding of most of the smaller crystallites when placed in water. The initial decrease was followed by a constant degree of crystallinity during which the unfolding of larger crystals took place. This continued until complete unfolding and subsequent dissolution of the gel in water. Some important findings were that an increase in the molecular weight of PVA resulted in a smaller number of crystallites, but larger lamellar thicknesses. So although there were fewer crystals in such a system, the crystals formed were more stable. This analysis can be applied to the issue of stability when considering other semicrystalline PVA systems such as freeze/thawed hydrogels.

A mathematical model was also proposed by Mallapragada and Peppas [87] to describe the dissolution of semicrystalline polymers. The development of this model was based on free energy changes during the crystal unfolding and disentanglement process in order to predict the overall dissolution kinetics. In addition, further work [88] examined controlled release systems based on semicrystalline polymers that exhibit such dissolution behavior in the presence of a solvent. Drug release was found to be controlled by the rate of crystal dissolution in a solvent (water or a biological fluid). A model was also developed to predict the drug release from such a device. The new system was described as a crystal dissolution-controlled system, a type of phase-erosion polymeric system.

In addition, such semicrystalline materials of PVA were further examined in some rather unique ways by Mallapragada et al. [89, 90]. Specifically, the crystal dissolution-controlled system described was investigated for the controlled release of metronidazole [89]. The crystalline phase of this system was also characterized more closely in terms of modifications of drug-release profiles [90].

Narasimhan et al. [91, 92] also examined the dissolution of PVA in water using magnetic resonance imaging to detect changes in the microstructure and molecular mobility. In this work, however, PVA samples were prepared by a quenching technique that yielded samples with degrees of crystallinity less than 5%. Such imaging techniques gave information as to solvent concentration profiles and the spatial variation of the water self-diffusion coefficient. This analysis shows promise for assisting in the development of drug-delivery systems based on polymer dissolution.

5
PVA Blends by Conventional and Freezing/Thawing Techniques

There has been a considerable amount of research which has investigated the addition of other components, notably other polymers, with poly(vinyl alcohol) to create a material with desirable characteristics and properties. This section will discuss the work involving pH-sensitive polymeric systems containing PVA and poly(acrylic acid) (PAA). The addition of poly(ethylene glycol) (PEG) into a poly(vinyl alcohol) system to reduce protein adsorption will also be discussed.

pH-sensitive membranes of interpenetrating polymer networks of PVA and PAA were investigated by Gudeman and Peppas [93, 94]. Membranes were prepared by chemical crosslinking techniques to contain varying degrees of crosslinking and ionic content. The molecular weight between crosslinks was found to vary from 270 to 40,000 from equilibrium swelling studies. The mesh size, upon swelling, varied from 19 to 710 Å as shown in Fig. 9. The oscillatory swelling behavior of the system was investigated over the pH range 3–6. An increase of 13–86% in the mesh size was observed when going from pH 3 to 6. The swelling ratio was also found to increase with a decrease in the ionic strength of the swelling environment. Swelling and syneresis occurred more abruptly with more loosely crosslinked membranes. Permeation studies indicated that solute separation was greatly affected by size exclusion and ionic interactions. These results showed that desired separation and selectivity characteristics can be obtained by appropriately designing the gel.

The diffusion of solutes in composite membranes of PVA and PAA was studied by Hickey and Peppas [95, 96]. These membranes were prepared by the freezing and thawing of aqueous solutions of the two components. The membranes were found to be strong and swelled to equilibrium within a few hours

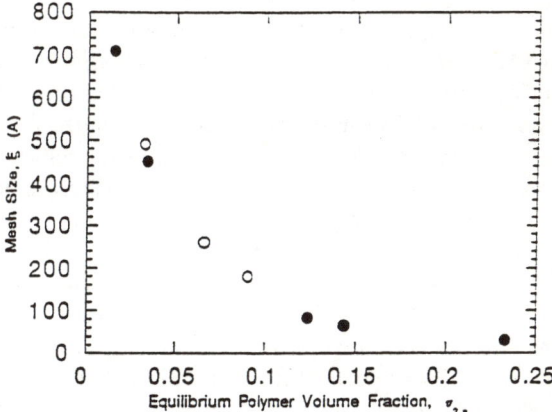

Fig. 9. Mesh size, ξ, as a function of the equilibrium polymer volume fraction, $\upsilon_{2,s}$, for PVA/PAA IPN membranes with PAA content of 50 (●) and 60 mol% (○)

at 25 °C. As the PAA content increased, the membrane swelling ratio increased due to the ionization of the carboxylic groups of PAA. In general, it was observed that the equilibrium mesh size was larger than that of PVA membranes prepared by the same techniques due to the ionic nature of PAA. Results from permeation studies with theophylline and FITC-dextran indicated that solute transport through such PVA/PAA membranes produced by freezing/thawing techniques is considerably slower than through chemically crosslinked membranes. A size exclusion phenomenon was observed and attributed to the physical network created by the formation of crystallites.

Solute transport in PVA/PAA interpenetrating networks was further investigated by Peppas and Wright [97–99] to address the possible binding of solutes during the transport process. Chemical crosslinking was implemented in the preparation of such networks. The diffusion of theophylline, vitamin B_{12}, and myoglobin was analyzed under varying pH conditions of 3 and 6. ATR-FTIR spectroscopy was used to qualitatively analyze the interactions between the ionized membranes and ionized solutes. It was found that solute binding occurred only for hydrogels in the ionized state.

Argade and Peppas [100] examined copolymers of PVA and PAA for their superabsorbent properties. A functionalized PVA was copolymerized with acrylic acid to prepare such materials. Additionally, the degradation of the PVA links between PAA chains was examined to determine the possible biodegradation to lower molecular weight polyacrylates.

A temperature- and pH-sensitive PVA/PAA interpenetrating polymer network was investigated by Shin et al. [101]. The release of the model drug, indomethacin, was studied under varying environmental conditions. The amount of drug released increased as the temperature was increased due to the dissociation of hydrogen bonding and resulting increase in swelling. However, a more significant change in drug release was observed when alternating the pH of the environment. This was attributed to the ionization created within the network. Overall, it was concluded that the release rate of a drug could be optimized in such a system by controlling the degree of swelling/deswelling as a function of pH and/or temperature. Further work by this group [102] examined the permeation of non-ionic and ionic solutes through interpenetrating polymer networks of PVA and PAA. Non-ionic solutes showed permeation results corresponding with the swelling in response to changes in the environment. However, with the permeation of ionic solutes, different trends were observed due to the attraction or repulsion between ionized groups in the gel and ionized electrolytes in solution.

Of additional interest is the incorporation of poly(ethylene glycol) (PEG) onto a poly(vinyl alcohol) hydrogel in order to decrease the adsorption of proteins and deposition of cells. Llanos and Sefton [103] examined the immobilization of PEG onto a PVA hydrogel using acetal or urethane linkages. Two mechanisms were investigated to accomplish this. In the first mechanism, the hydroxyl terminus of PEG was oxidized to an aldehyde. In the second mechanism, PEG was modified with a diisocyanate to produce an isocyanate end group. Upon completion of either mechanism, the modified PEG was grafted through a reaction with the hydroxyl groups of PVA. These grafted gels were characterized by various spectroscopic techniques. It was suggested that the immobilization was limited by the degree of mixing of PVA and PEG, not by the reactivity of the PEG end groups. Both mechanisms, therefore, were equally satisfactory resulting in a maximum oxyethylene content of between 70 and 75%. The findings were significant in that as the molecular weight of PEG was increased, it was expected to have fewer PEG molecules covalently bound to the surface. Thus, the biological properties of the surface would be adversely affected.

To gain a better understanding of the preparation of blends containing both PVA and PEG, there has been extensive work to further characterize polymer–polymer interactions. Sawatari [104] examined blends of PVA and poly(ethylene oxide) (PEO). In particular, different preparation conditions were investigated including the ratio of PVA to PEO, solvent concentration, and temperature. An increase in the amount of solvent resulted in a decrease in the crystallinity of the resulting films of the PVA/PEO blend. Inamura [105] more accurately described the phase separation of a system containing PVA, PEG, and water. It was found that the PEG molecular weight was a key issue in both phase separation and gelation. Further work by Inamura et al. [106] utilized viscometry to study interactions of the PVA-PEG-water system particularly between PVA and PEG. This work showed that there are repulsive interactions between the polymers and considerable incompatibility.

Recently, Masaro and Zhu [107] studied the diffusion of PEG in PVA solutions using pulsed-gradient NMR spectroscopy. In fact, the self-diffusion coefficients of ethylene glycol, its oligomers, and polymers were determined in aqueous solutions and gels of PVA. A diffusion model was proposed to describe the effect of diffusant size, polymer concentration, and temperature. Continued work in this area is focusing on more accurately predicting the effect of the diffusant shape and molecular interactions.

6
Biomedical and Pharmaceutical Applications of PVA

PVA hydrogels have been used for numerous biomedical and pharmaceutical applications [32]. PVA hydrogels have certain advantages that make them excellent candidates for biomaterials. Some of these advantages include their non-toxic, non-carcinogenic, and bioadhesive characteristics, as well as their associated ease of processing. PVA has an uncomplicated chemical structure and modifications are possible by simple chemical reactions. In addition, PVA gels exhibit a high degree of swelling in water (or biological fluids) and a rubbery and elastic nature. Because of these properties, PVA is capable of simulating natural tissue and can be readily accepted into the body. PVA gels have been used for contact lenses, the lining for artificial hearts, and drug-delivery applications. In addition, they have been shown to have potential applications for soft tissue replacements, articular cartilage, catheters, artificial skin, artificial pancreas, and hemodialysis membranes. Some specific applications of PVA will be discussed here.

Peppas and Merrill [108, 109] conducted some of the earliest work in considering PVA hydrogels as biomaterials. PVA gels were examined for their use in applications where blood compatibility was a major issue. The physical and surface properties of the gels were examined. In particular, the heparinization of PVA to provide for biocompatibility was examined. In addition, the elastic nature of such PVA gels was investigated and recognized for its importance in various biomedical applications. Based on such observations, PVA hydrogels were investigated for the possible reconstruction of vocal cords [110, 111]. Crosslinked PVA hydrogels have also been considered as candidates for biomembranes in artificial kidney applications [112].

Much research has focused on biocompatibility issues to more accurately ascertain information as to the use of such PVA gels for biomedical or pharmaceutical applications. Tamura et al. [113] examined the use of PVA gels prepared by freezing/thawing processes as a material for medical use. The gel had a water content of 80 to 90% by weight, high mechanical strength, and rubber-like elasticity. The characteristics of the gel, which resembled those of natural tissue, did not change after long-term implantation. In addition, gel samples that were implanted subcutaneously or intramuscularly into rabbits showed good bioinertness. No adhesion to surrounding tissue was found. Although there was some infiltration of inflammatory cells initially, it disappeared during the second week after implantation. They concluded that the material would be useful for clinical applications.

Further biocompatibility issues of PVA were addressed by Fujimoto et al. [114]. This work focused on PVA gels that had been annealed in the presence of glycerol. When such materials were examined for their interactions with blood components, reduced adsorption and platelet adhesion were observed due to the addition of glycerol. Glycerol essentially altered the surface of the PVA gel. They described the mechanism as being due to increased tethered PVA chains on the surface which served to decrease the direct contact of blood components with the surface.

There has been a great deal of work focusing on the use of hydrogel materials for the development of articular cartilage. Oka et al. [115] investigated PVA hydrogels for such applications. They reported on the biocompatibility as well as the mechanical properties of PVA gels in relation to their usefulness as artificial articular cartilage. They examined such aspects as lubrication, load bearing, biocompatibility, and attachment of the material to the bone to look at the overall biomechanics of the material. Problems that were encountered with the use of PVA for such applications included its wear resistance properties. However, continued work has examined the use of higher molecular weight PVA along with a new annealing process to enhance the mechanical properties.

With the continuation of this work by Noguchi et al. [116], the new manufacturing process was found to increase the tensile strength to 17 MPa, which resembles that of normal human articular cartilage. However, it was necessary to investigate the biocompatibility of the improved PVA hydrogel. A series of in vivo tests was performed in an intra-articular environment. It was found that upon implantation of the material into the rabbit knee joint, there was only slight to mild inflammation initially. This new material was found to have excellent biocompatibility and physical properties for use as artificial articular cartilage. They did, however, report some issues that needed to be addressed such as degradation in vivo, the reaction of tissue to wear particles, and methods for shaping the material.

In considering such applications, the mechanical and wear properties of the material of interest are critical. Cha et al. [117] made additional contributions in preparing PVA gels with increased mechanical strength and hardness. The method involved the preparation and characterization of gels of low water content by low temperature crystallization of PVA in a mixed solvent of dimethyl sulfoxide (DMSO) and water.

Further biomedical applications of PVA were examined by Scotchford et al. [118]. In this work, blends of collagen and PVA were cast into films and used as substrates for the culture of osteoblasts. They investigated the ratio of collagen to PVA as well as the crosslinking technique. The crosslinking technique of interest, dehydrothermal treatment, involved two dehydration steps followed by a crosslinking step by annealing. This method was found to be favored over chemical crosslinking (through the use of glutaraldehyde) due to increased biological stability. Overall, this work showed the potential of PVA for orthopedic applications.

PVA gels have also been closely investigated in terms of their diffusive characteristics. The diffusion of macromolecules through crosslinked PVA networks was studied by Sorensen and Peppas [119]. Such an analysis was important particularly when considering such materials for potential drug-delivery applications. Further research by Peppas and collaborators [120, 121] more closely analyzed drug and protein diffusion in such PVA materials. Specifically, the release of a water-soluble drug, theophylline, was investigated from matrices of crosslinked PVA. The effect of the molecular weight between crosslinks, $\overline{M}_c$, on the diffusion of bovine serum albumin (BSA) was also examined as shown in Fig. 10. The diffusion coefficient of BSA was normalized with respect to its diffusion coefficient in pure water at 37 °C. Additional work by Gander et al. [122–124] further demonstrated the appropriateness of crosslinked PVA for various pharmaceutical applications. Kim and Lee [125] also investigated PVA gels for oral drug-delivery applications. They prepared spherical PVA beads by suspension polymerization techniques. The effect of the crosslinking density on the overall swelling and drug-release behavior was examined.

Additional drug-release applications have been investigated which employ the bioadhesive nature of PVA gels. Morimoto et al. [126, 127] examined the

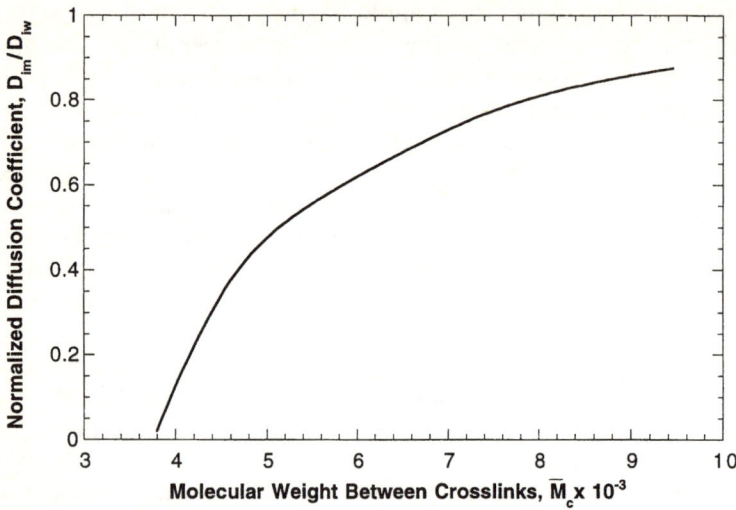

Fig. 10. Normalized diffusion coefficient of bovine serum albumin as a function of molecular weight between crosslinks for PVA hydrogels swollen to equilibrium

controlled release of several drugs from crosslinked PVA gel carriers for rectal administration. Of specific interest was the transrectal delivery of drugs for the treatment of hypertension. Buccal delivery systems of PVA hydrogels were also examined by Tsutsumi et al. [128] for the release of ergotamine tartrate for the treatment of migraine headaches. Recently, there has been considerable work in our own laboratory with the use of freeze/thawed PVA gels for various bioadhesive applications [76–82]. This work will be discussed in more detail in the following section.

PVA gels have shown promise for use as a soft contact lens material. Peppas and Yang [129] investigated the transport of oxygen through crosslinked and pure PVA materials. In particular, the permeation of oxygen was related to the overall polymer structure. Cha et al. [130] performed preliminary studies with transparent PVA gels that were prepared using low temperature crystallization from PVA solutions in a mixed solvent of water and DMSO. Such gels showed the potential for contact lens applications due to a decrease in protein adsorption. Hyon et al. [131] further developed such materials with certain enhanced properties for contact lens applications using low temperature crystallization with a water-miscible organic solvent. Such a technique allowed for the preparation of a transparent gel with high mechanical strength and high water content. This material was further investigated as a soft contact lens material through numerous experiments. It was determined that the gel had five times the tensile strength of poly(2-hydroxyethyl methacrylate), the most widely used soft lens material today. In addition, the adsorption of proteins to the PVA material was half to one thirtieth of the adsorption to conventional lens materials. Additional studies showed that the corneal epithelium of rabbit eyes was not affected by wearing the lens material.

Kuriaki et al. [132] also examined PVA gels with potential contact lens applications. They reported a process for preparing transparent PVA gels with high tensile strength. PVA solutions of 20 wt % were freeze-dried at –4 °C. Dried PVA gels were then swollen in an ethanol and water solution followed by the exchange of water. The resulting transparent gels had high tensile strength and elongation.

Hirai and collaborators [133–136] have also examined some rather interesting applications of PVA gels involving shape-memory properties. The contraction and relaxation of chemically crosslinked PVA gels in response to the exchange of the solvents, DMSO and water, was examined. The research progressed further with the application of an electric field to actuate a rapid gel response. PVA gels prepared by freezing and thawing techniques were also examined for their shape-restoring properties. In fact, chemical crosslinks were introduced into such freeze/thawed gels to enhance the shape-memory characteristics. It was proposed that the added chemical crosslinks served to "remember" the distribution of physical crosslinks within the gel to restore the shape upon elongation.

PVA gels have also been proposed for waste water cleaning applications. Okazaki et al. [137] investigated the immobilization of microorganisms on PVA gel particles as such a method for waste water treatment and compared it with the use of poly(ethylene glycol) (PEG) and polyacrylamide (PAAm) gels. PVA hydrogel particles were prepared by freezing and thawing techniques. The material was found to have a treating capacity for synthetic sewage of two to three times that of standard activated sludge processes. When compared with PEG and PAAm networks, PVA gels have the presence of voids of several microns which were formed upon freezing. Therefore, the PVA gel had many more desirable characteristics, such as a large free volume of water and good oxygen permeability. This created a favorable environment for microorganisms. An effluent treatment was created with microorganisms immobilized in the frozen PVA gel. This research has shown to be an interesting as well as a promising application of PVA gels.

7
Summary

Poly(vinyl alcohol) has been characterized on many levels and examined for numerous applications. It is a polymer of great interest because of its relatively simple chemical structure, ease of processing, and potential use in pharmaceutical and biomedical fields. The crystalline nature of PVA has been of specific interest. Crystallinity in PVA gels has been introduced by various methods including annealing, aging, low temperature crystallization, and repeated cycles of freezing and thawing. Combinations of the above techniques to enhance certain properties have also been studied. In addition, various solvents have been introduced during preparation techniques to improve certain properties of PVA gels, specifically with freeze/thawed gels. In particular, preparing transparent gels with high mechanical integrity has been of great interest particularly for contact lens applications. Overall, PVA gels prepared by freezing and thawing methods have shown many improved properties over traditional PVA gels prepared by chemical crosslinking techniques with increased mechanical strength being the most notable.

PVA hydrogels prepared by freezing and thawing processes have shown incredible potential for applications in medicine and pharmacy. However, the long-term stability of such gels still remains an important issue. In addition, there are numerous parameters that can impact the resulting structure and properties of the prepared gels. Initial insight has been obtained by several researchers over the past 20 to 25 years. However, there is still a need to gain a better understanding of structure-property relationships with specific emphasis on biomedical and pharmaceutical applications.

Acknowledgments. This work was supported in part by grants from the National Institutes of Health (GM 43337) and the Showalter Foundation.

References

1. Tubbs RK (1966) J Polym Sci A1 4:623
2. Finch CA (1973) Poly(vinyl alcohol): properties and applications. Wiley, New York
3. Peppas NA (1987) Hydrogels in medicine and pharmacy, vol 1, Polymers. CRC Press, Boca Raton, FL
4. Peppas NA (1987). In: Peppas, NA (ed) Hydrogels in medicine and pharmacy, vol 2, Polymers. CRC, Boca Raton, FL, pp 1–48.
5. Danno A (1958) J Phys Soc Jpn 13:722
6. Saito O (1959) J Phys Soc Jpn 14:792
7. Dieu J, Desreux V (1959) Large Radiation Sources in Industry 1:341
8. Sakurada I, Mori N (1959) Seni Gakkaishi 15:948
9. Peppas NA, Merrill EW (1976) J Appl Polym Sci 20:1457
10. Bray JC, Merrill EW (1973) J Appl Polym Sci 17:3781
11. Peppas NA, Merrill EW (1974) Techn Chron 43:559
12. Peppas NA, Merrill EW (1976) J Polym Sci Polym Chem Ed 14:441
13. Stauffer SR, Peppas NA (1992) Polymer 33:3932
14. Bunn CW (1948) Nature 161:929
15. Flory PJ (1953) Principles of polymer chemistry. Cornell University Press, Ithaca, NY
16. Keller A (1957) Philos Mag 2:1171
17. Mandelkern L (1967) J Polym Sci C18:51
18. Harris HE, Kenney JF, Willcockson GW, Chiang, R, Friedlander, HN (1966) J Polym Sci A1 4:665
19. Fujii K (1973) In: Finch, CA (ed) PVA-properties and applications. Wiley, New York, pp 203–231
20. Peppas NA (1976) Eur Polym J 12:495
21. Peppas NA (1976) Org Coat Plast Chem Prepr 36:541
22. Peppas NA (1977) In: Labana SS (ed) Chemistry and properties of crosslinked polymers. Academic Press, New York, pp 469–478

23. Peppas NA, Hansen PJ (1982) J Appl Polym Sci 27:4787
24. Mallapragada SK, Peppas NA (1996) J Polym Sci 34:1339
25. Yamaura K, Suzuki M, Yamamoto M, Shimada R, Tanigami T (1995) J Appl Polym Sci 58:1787
26. Yamaura K, Kitahara H, Tanigami T (1997) J Appl Polym Sci 64:1283
27. Yamaura K, Ideguchi S, Tanigami, T (1998) J Appl Polym Sci 70:1661
28. Sone Y, Hirabayashi, K, Sakurada I (1953) Kobunshi Kagaku 10:7
29. Tanigami T, Nakashima Y, Murase K, Suzuki H, Yamaura K, Matsuzawa S (1995) J Mater Sci 30:5110
30. Tanigami T, Yano K, Yamaura K, Matsuzawa S (1995) Polymer 36:2941
31. Tanigami T, Murase K, Yamaura K, Matsuzawa S (1994) Polymer 35:2573
32. Tanigami T, Yano K, Yamaura K, Matsuzawa S (1995) Polymer 36:2941
33. Sakurada I, Nukushina Y, Sone Y (1955) Kobunshi Kagaku 12:510
34. Tadokoro H, Nagai J, Seki S, Nitta I (1961) Bull Chem Soc Jpn 34:1504
35. Peppas NA (1977) Makromol Chem 178:595
36. Peppas NA, Stauffer SR (1991) J Controlled Release 16:305
37. Lozinsky VI (1998) Russian Chem Rev 67:573
38. Peppas NA (1975) Makromol Chem 176:3433
39. Nambu M (1984) US Patent 4,472,542
40. Nishinari K, Watase M, Ogino K, Nambu M (1983) Polymer Commun 24:345
41. Mano I, Goshima H, Nambu M, Iio M (1986) Magn Reson Med 3:921–926
42. Yokoyama F, Masada I, Shimamura K, Ikawa T, Monobe K (1986) Colloid Polym Sci 264:595
43. Watase M, Nishinari K (1985) J Appl Polym Sci Polym Phys Ed 23:1803
44. Watase M, Nishinari K (1989) Makromol Chem 190:155
45. Hyon SH, Ikada Y (1987) US Patent 4,663,358
46. Hyon SH, Cha WI, Ikada Y (1989) Kobunshi Ronbunshu 46:673
47. Berghmans H, Stoks W (1986) In: Kleintjes LA, Lemstra, PJ (eds) Integration of fundamental polymer science and technology. Elsevier, London, pp 218–229
48. Lozinsky VI, Domotenko LV, Vainerman ES, Mamtsis AM, Rogozhin SV (1986) Polym Bull 15:333
49. Lozinsky VI, Zubov AL, Kulakova VK, Titova EF, Rogozhin SV (1992) J Appl Polym Sci 44:1423
50. Lozinsky VI, Solodova EV, Zubov AL, Simenel IA (1995) J Appl Polym Sci 58:171
51. Lozinsky VI, Domotenko LV, Zubov AL, Simenel IA (1996) J Appl Polym Sci 61:1991
52. Urushizaki F, Yamaguchi H, Mizumachi H (1986) Yakugaku Zasshi 106:491
53. Urushizaki F, Yamaguchi H, Nakamura K, Numajiri S, Sugibayashi K, Morimoto Y (1990) Int J Pharm 58:135
54. Ohkura M, Kanaya T, Kaji K (1992) Polymer 33:3686
55. Ohkura M, Kanaya T, Kaji K (1992) Polymer 33:5044
56. Kanaya T, Ohkura M, Kaji K (1994) Macromolecules 27:5609
57. Kanaya T, Ohkura M, Takeshita H, Kaji K (1995) Macromolecules 28:3168
58. Takeshita H, Kanaya T, Nishida K, Kaji K, Imai, M (1995) Polym Prepr Jpn (Engl Ed) 44:139
59. Hatakeyama T, Yamauchi A, Hatakeyama H (1987) Eur Polym J 23:361
60. Cha WI, Hyon SH, Ikada Y (1993) Makromol Chem 194:2433
61. Murase K, Tanigami T, Yamaura K, Matsuzawa S (1991) Polym Prepr Jpn (Engl Ed) 40:952
62. Trieu HH, Qutubuddin S (1994) Colloid Polym Sci 272:301
63. Tazaki M, Kida T, Homma T (1989) Kobunshi Ronbunshu 46:733
64. Suzuki M, Matsuzawa M, Saito K (1991) Polym Prepr Jpn (Engl Ed) 40:953
65. Suzuki M, Tateishi T, Matsuzawa M, Saito K (1996) Macromol Symp 109:55

66. Bao QB, Higham PA (1991) US Patent 5,047,055
67. Takamura A, Arai M, Ishii F (1987) Yakugaku Zasshi 107:233
68. Takamura A, Ishii, F (1991) Yakugaku Zasshi 111:45
69. Takamura A, Ishii F, Hidaka H (1992) J Controlled Release 20:21
70. Hyon SH, Cha WI, Oka M, Ikada Y (1993) Polym Prepr Jpn (Engl Ed) 42
71. Peppas NA (1996) Proc Topical Conf Process Structure Prep Polym Materials 1:321
72. Peppas NA, Scott JE (1992) J Controlled Release 18:95
73. Ficek BJ, Peppas NA (1993) J Controlled Release 27:259
74. Ficek BJ, Peppas NA (1994) In: Mikos AG, Murphy R, Bernstein H, Peppas NA (eds) Biomaterials for drug and cell delivery. Materials Research Society, Pittsburgh, PA, pp 223–226.
75. Hickey AS, Peppas NA (1995) J Membr Sci 107:229
76. Peppas NA, Mongia NK (1997 Eur J Pharm Biopharm 43:51
77. Peppas NA, Mongia N, Luttrell AS (1995) Proc World Meeting APGI/APV 1:817
78. Peppas NA, Anseth KS, Mongia NK (1996) Trans World Biomat Congress 5:643
79. Mongia NK, Anseth KS, Peppas NA (1996) J Biomat Sci Polym Ed 7:1055
80. Luttrell AS, Mongia NK, Peppas NA (1994) Abstr AIChE Meeting 202c
81. Peppas NA, Borcherding ASL (1996) Proc Int Symp Control Rel Bioact Mater 23:145–146
82. Peppas NA, Mongia NK, Bugert CA (1996) Proc Int Symp Control Rel Bioact Mater 23:157
83. Hassan CM, Peppas NA (1998) Polym Mater Sci Eng Proc 79:473
84. Hassan CM, Trakarnpan P, Peppas NA (1998) In: Amjad Z (ed) Water soluble polymers: solution properties and applications. Plenum Press, New York, pp 31–40
85. Hassan CM, Peppas NA (1998) Proc Int Symp Control Rel Bioact Mater 25:50
86. Mallapragada SK, Peppas NA (1995) Polym Mater Sci Eng Proc 73:22
87. Mallapragada SK, Peppas NA (1997) AIChE J 43:870
88. Mallapragada SK, Peppas NA (1997) J Controlled Release 45:87
89. Mallapragada SK, Peppas NA, Colombo P (1996) Polym Mater Sci Engin Proc 74:416
90. Mallapragada SK, Peppas NA, Colombo P (1997) J Biomed Mater Res 36:125
91. Narasimhan B, Snaar JEM, Bowtell RW, Morgan S, Melia CD, Peppas NA (1999) Macromolecules 32:704
92. Snaar JEM, Bowtell R, Melia CD, Morgan S, Narasimhan B, Peppas (1998) Magnetic Resonance Imaging 16:691
93. Gudeman LF, Peppas NA (1995) J Membr Sci 107:239
94. Gudeman LF, Peppas NA (1995) J Appl Polym Sci 55:919
95. Hickey AS, Peppas NA (1997) Polymer 38:5931
96. Peppas NA, Hickey AS (1997) In: Peppas NA, Mooney DJ, Mikos AG, Brannon Peppas L (eds) Biomaterials: carriers for drug delivery and scaffolds for tissue engineering. AIChE, New York, pp 328–330
97. Peppas NA, Wright SL (1996) Macromolecules 29:8798
98. Peppas NA, Wright SL (1996) Report Poval Committee 108:96
99. Peppas NA, Wright SL (1998) Eur J Pharm and Biopharm 46:15
100. Argade AB, Peppas NA (1998) J Appl Polym Sci 70:817
101. Shin HS, Kim SY, Lee YM (1997) J Appl Polym Sci 65:685
102. Shin HS, Kim SY, Lee YM, Lee KH, Kim SJ, Rogers CE (1998) J Appl Polym Sci 69:479
103. Llanos GR, Sefton MV (1991) Macromolecules 24:6066
104. Sawatari C (1996) Toyota Kenkyu Hokoku 49:77
105. Inamura I (1986) Polym J 18:269
106. Inamura I, Akiyama K, Kubo Y (1997) Polym J 29:119
107. Masaro L, Zhu XX (1998) Macromolecules 31:3880

108. Peppas NA, Merrill EW (1977) J Biomed Mater Res 11:423
109. Peppas NA, Merrill EW (1977) J Appl Polym Sci 21:1763
110. Peppas NA, Benner RE Jr, Sorensen, RA (1979) Proc IUPAC 26:1539
111. Peppas NA, Benner RE Jr (1980) Biomaterials 1:158
112. Peppas NA (1977) Polym Prepr 18(1):794
113. Tamura K, Ike O, Hitomi S, Isobe J, Shimizu Y, Nambu M (1986) Trans Amer Soc Artif Organs 32:605
114. Fujimoto K, Minato M, Ikada Y (1994) In: Shalaby SW, Ikada Y, Langer R, Williams J (eds) ACS Symposium Series 540, Polymers of biological and biomedical significance. Plenum Press, New York, pp 228–241
115. Oka M, Naguchi T, Kumar P, Ikeuchi K, Yamamuno T, Hyon SH, Ikada Y (1990) Clinical Materials 6:361
116. Noguchi T, Yamamuro T, Oka M, Kumar P, Kotoura Y, Hyon SH, Ikada Y (1991) J Appl Biomater 2:101
117. Cha WI, Hyon SH, Ikada Y (1996) Macromol Symp 109:115
118. Scotchford CA, Cascone MG, Downes S, Giusti P (1998) Biomaterials 19:1
119. Sorensen RA, Peppas NA (1979) Proc IUPAC 26:1108
120. Reinhart CT, Korsmeyer RW, Peppas NA (1981) Int J Pharm Techn 2(2):9
121. Korsmeyer RW, Peppas NA (1981) J Membr Sci 9:211
122. Gander B, Gurny R, Doelker E, Peppas NA (1983) Proc Int Symp Control Rel Pharm 5
123. Gander B, Gurny R, Doelker E, Peppas NA (1989) Pharm Res 6:578
124. Gander B, Gurny R, Doelker E, Peppas NA (1988) Proc Int Symp Contr Rel Bioactiv Mater 15:109
125. Kim CJ, Lee (1990) Polym Mater Sci Eng Proc 63:64
126. Morimoto K, Nagayasu A, Fukanoki S, Morisaka K, Hyon SH, Ikada Y (1989) Pharm Res 6:338
127. Morimoto K, Fukanoki S, Morisaka K, Hyon SH, Ikada, Y (1989) Chem Pharm Bull 37:2491
128. Tsutsumi K, Takayama K, Machida Y, Ebert CD, Nakatomi I, Nagai T (1994) STP Pharma Sci 4:230
129. Peppas NA, Yang WH (1980) Proc IUPAC 27(4):28
130. Cha WI, Hyon SH, Ikada Y (1991) Kobunshi Ronbunshu 48:425
131. Hyon SH, Cha WI, Ikada Y, Kita M, Ogura Y, Honda Y (1994) J Biomater Sci Polymer Edn 5:397
132. Kuriaki M, Nakamura K, Mizutani J (1989) Kobunshi Ronbunshu 46:739
133. Hirai T, Maruyama H, Nemoto H, Suzuki T, Hayashi S (1990) Polym Prepr Jpn (Engl Ed) 39:942
134. Hirai T, Nemoto H, Suzuki T, Hayashi S (1991) Polym Prepr Jpn (Engl Ed) 40:1297
135. Hirai T, Maruyama H, Suzuki T, Hayashi S (1992) J Appl Polym Sci 46:1449
136. Hirai T, Maruyama H, Suzuki T, Hayashi S (1992) J Appl Polym Sci 45:1849
137. Okazaki M, Hamada T, Fujii H, Mizobe A, Matsuzawa S (1995) J Appl Polym Sci 58:2235

Editor: Prof. K. Dušek
Received: June 1999

Applications of 1,1-Diphenylethylene Chemistry in Anionic Synthesis of Polymers with Controlled Structures

Roderic P. Quirk[1], Taejun Yoo[1], Youngjoon Lee[1], Jungahn Kim[2], Bumjae Lee[3]

[1] Maurice Morton Institute of Polymer Science, The University of Akron, Akron, Ohio 44325-3909, USA, *e-mail: Quirk@polymer.uakron.edu*
[2] Division of Polymers, Korea Institute of Science and Technology, P. O. Box Cheongryang 131, Seoul, Korea
[3] Department of Fine Chemicals Engineering and Chemistry, Chungnam National University, Taejon, 305-764, Korea

The use of 1,1-diphenylethylenes in anionic polymerization is reviewed. The structure and reactivity of 1,1-diphenylethylenes and the corresponding simple and polymeric 1,1-diphenylalkyllithiums are described. The applications of 1,1-diphenylethylene chemistry include:
1. Their use to form initiators
2. End-capping agents to attenuate reactivity for crossover to reactive, polar monomers or for functionalization reactions
3. For chain-end and in-chain functionalization using substituted 1,1-diphenylethylenes and 1,1-diarylethylenes
4. The preparation and applications of non-homopolymerizable 1,1-diphenylethylene-functionalized macromonomers
5. The use of bis(1,1-diphenylethylenes) and tris(1,1-diphenylethylenes) as precursors for hydrocarbon-soluble dilithium and trilithium initiators, respectively
6. The use of bis(1,1-diphenylethylenes) and tris(1,1-diphenylethylenes) as living linking agents to prepare heteroarm (miktoarm) star-branched polymers

Keywords. Anionic polymerization, Living anionic polymerization, 1,1-Diphenylalkyllithiums, Functionalized polymers, Block copolymers, Macromonomers, Star-branched polymers, Dilithium initiators, Trilithium initiators, Multifunctional initiators, Living linking reactions, Heteroarm star polymers, Miktoarm star polymers

1	Introduction	69
2	Reactions of Alkyllithium Compounds with 1,1-Diphenylethylene	70
2.1	Stoichiometry, Kinetics, and Mechanism.	70
2.1.1	Stoichiometry	70
2.1.2	Kinetics and Mechanism	72
2.2	Stability and Structure of 1,1-Diphenylalkyllithiums.	75
2.3	Initiators for Anionic Polymerization.	79
2.3.1	Alkyl Methacrylates	79
2.3.2	Styrenes and Dienes.	80

3	**Reactions of Polymeric Organolithium Compounds with 1,1-Diphenylethylene**	84
3.1	Kinetics	84
3.2	Poly(styryl)lithium.	85
3.2.1	Substituent Effects	86
3.2.2	Synthetic Utility	89
3.3	Poly(dienyl)lithiums	90
3.3.1	Synthetic Utility	92
4	**Polymeric 1,1-Diphenylalkyllithium Initiators for Block Copolymer Synthesis**	93
4.1	Crossover Reactions of Polymeric 1,1-Diphenylalkyllithiums to Dienes	94
4.2	Crossover Reactions of Polymeric 1,1-Diphenylalkyllithiums to Styrene	95
5	**Copolymerization of 1,1-Diphenylethylene with Styrenes and Dienes**	96
5.1	Copolymerization with Styrenes	97
5.2	Copolymerization with Dienes	99
6	**Anionic Functionalization Chemistry of 1,1-Diphenylethylenes** .	102
6.1	Introduction to Anionic Functionalization Reactions.	102
6.2	Applications of 1,1-Diphenylethylene in Specific Functionalization Reactions	102
6.2.1	Carbonation.	102
6.2.2	Sulfonation	103
6.2.3	Epoxide-Functionalized Macromonomers	104
6.3	General Aspects of Anionic Functionalization Chemistry with 1,1-Diphenylethylenes	105
6.3.1	Protecting Groups	107
6.3.2	Phenol Functionality.	108
6.3.3	Tertiary Amine Functionality	109
6.3.4	Primary Amine Functionality	112
6.3.5	Carboxyl Functionality	114
6.3.6	Condensation Macromonomers: Diphenol Functionality	115
6.3.7	Trimethylsilyl Group Labeling for Molecular Weight Measurements	117
6.3.8	Fluorescent Group Labeling.	118
6.3.9	Copolymerization with Substituted 1,1-Diphenylethylenes	121
6.3.10	Conclusions Regarding 1,1-Diphenylethylene Functionalizations .	123

7	Preparation of 1,1-Diphenylethylene-Functionalized Macromonomers and their Applications for Synthesis of Heteroarm, Star-Branched Polymers	123
8	Difunctional and Trifunctional Organolithium Initiators Based on 1,1-Diphenylethylene	132
8.1	Difunctional Organolithium Initiators	132
8.1.1	1,4-Dilithio-1,1,4,4-tetraphenylbutane	133
8.1.2	1,3-Bis(1-phenylethenyl)benzene	133
8.2	Trifunctional Organolithium Initiators	139
9	Multifunctional Living Linking Agents Based on 1,1-Diphenylethylene	141
9.1	1,3-Bis(1-phenylethenyl)benzene	143
9.1.1	Preparation of Heteroarm, Star-Branched Polymers	145
9.1.1.1	Preparation of Heteroarm, Star-Branched Styrene-Butadiene Thermoplastic Elastomers	149
9.1.2	Preparation of Functionalized Cyclic Polymers	151
10	Conclusions	154
References		155

1
Introduction

One of the goals of synthetic polymer chemistry is to develop methods for the synthesis of polymers with predictable, well-defined structures. Living anionic polymerization, particularly alkyllithium-initiated polymerizations of styrene and diene monomers, provides one of the best methodologies for the synthesis of polymers with control of the major variables affecting polymer properties [1–11]. These variables include molecular weight, molecular weight distribution, copolymer composition and microstructure, tacticity, chain-end and/or in-chain functionality, architecture, and morphology. Thus, for polymerizations with monomer/initiator/solvent systems that proceed in the absence of chain termination and chain transfer reactions, polymers can be prepared with low degrees of compositional heterogeneity [3, 7]. However, the synthesis of such well-defined polymers requires careful experimental procedures [12–14]; unfortunately, overreliance on the potential of the method has fostered the erroneous belief that the mere use of anionic polymerization obviates the necessity of characterization and structure proof [7]. This review will describe the unique chemistry and applications of 1,1-diarylethylenes in anionic polymer synthesis. The background, scope, and limitations of these procedures will be critically presented.

2
Reactions of Alkyllithium Compounds with 1,1-Diphenylethylene

2.1
Stoichiometry, Kinetics, and Mechanism

The discovery by Firestone scientists [15] that anionic polymerization of isoprene with lithium metal produced high *cis*-1,4-polyisoprene and the elucidation of the concept of living anionic polymerization by Szwarc and coworkers [16, 17] generated considerable research interest in elucidating the mechanistic details of anionic polymerization and related processes. It was soon found that the mechanism of initiation in anionic polymerization of vinyl monomers with alkyllithium compounds and other organometallic compounds is complicated by chain-end association and cross-association phenomena in hydrocarbon solvents [18] and by the presence of a variety of ionic species in polar media [4, 6, 9]. Consequently, several simple addition reactions of alkyllithium initiators to vinyl-type compounds were utilized as models for these initiation reactions without the added complication of concurrent propagation which occurs with most vinyl monomers. Thus, the kinetics and mechanism of the addition reactions of alkyllithium compounds with 1,1-diphenylethylene (DPE) were investigated as a model for the reaction of alkyllithium compounds with styrenes and dienes in the initiation step for anionic polymerization. The results of these investigations provide the foundation for understanding the many synthetic applications of DPE chemistry in anionic polymerization.

2.1.1
Stoichiometry

It is generally considered that polymeric and simple organolithium compounds react quantitatively and relatively rapidly with 1,1-diphenylethylene (DPE) to produce the corresponding 1,1-diphenylalkyllithium species, as shown in Eq. (1) [19]:

$$C_4H_9Li + CH_2{=}C\underset{C_6H_5}{\overset{C_6H_5}{|}} \longrightarrow C_4H_9CH_2\underset{C_6H_5}{\overset{C_6H_5}{\underset{|}{\overset{|}{C}}}}Li \tag{1}$$

The stoichiometric reaction of *n*-butyllithium with DPE was first examined by Ziegler and coworkers [20] in benzene solution at room temperature as shown in Eq. (2):

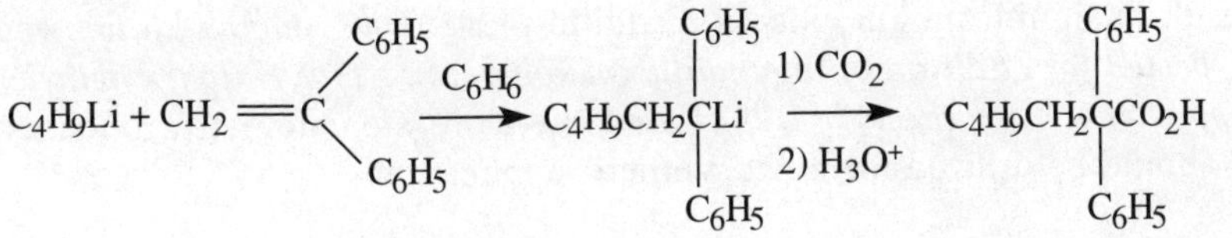

$$C_4H_9Li + CH_2=C\begin{matrix}C_6H_5\\ \\C_6H_5\end{matrix} \xrightarrow{C_6H_6} C_4H_9CH_2\underset{\underset{C_6H_5}{|}}{\overset{\overset{C_6H_5}{|}}{C}}Li \xrightarrow[2)\ H_3O^+]{1)\ CO_2} C_4H_9CH_2\underset{\underset{C_6H_5}{|}}{\overset{\overset{C_6H_5}{|}}{C}}CO_2H \quad (2)$$

The structure of the carboxylated derivative of the addition product was proven from a mixed melting point determination with an authentic sample. Köbrich and Stöber [21] reinvestigated this reaction in tetrahydrofuran (THF) at −80 to 20 °C and isolated α,α-diphenylheptanoic acid in 98% yield; protonation and alkylation of the intermediate 1,1-diphenylhexyllithium with water and n-butyl bromide formed 1,1-diphenylhexane and 5,5-diphenyldecane in 99% and 97% yields, respectively. However, Evans and George [22] have reported that further reversible addition can occur when a large molar excess (6.4-fold) of 1,1-diphenylethylene is reacted with n-butyllithium in benzene at 30 °C as shown in Scheme 1. The amount of 1,1,3,3-tetraphenyloctane isolated after hydrolysis was much less than the amount of 1,1-diphenylhexane; therefore it was concluded that the second equilibrium step in Scheme 1 strongly favors the monoadduct. No 1,1,3,3-tetraphenyloctane was detected when only a 1.8-fold excess of DPE was used [22, 23]. From the kinetics of the reaction it was concluded that the addition of n-butyllithium to DPE is irreversible [23].

The reaction of dibutylmagnesium with an unspecified amount of 1,1-diphenylethylene in hexamethylphosphoric triamide (HMPA) has been reported

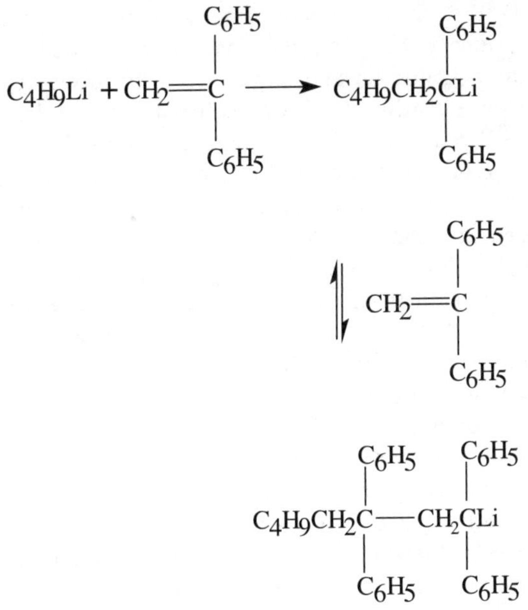

$$C_4H_9Li + CH_2=C\begin{matrix}C_6H_5\\ \\C_6H_5\end{matrix} \longrightarrow C_4H_9CH_2\underset{\underset{C_6H_5}{|}}{\overset{\overset{C_6H_5}{|}}{C}}Li$$

$$\Updownarrow \ CH_2=C\begin{matrix}C_6H_5\\ \\C_6H_5\end{matrix}$$

$$C_4H_9CH_2\underset{\underset{C_6H_5}{|}}{\overset{\overset{C_6H_5}{|}}{C}}-CH_2\underset{\underset{C_6H_5}{|}}{\overset{\overset{C_6H_5}{|}}{C}}Li$$

Scheme 1.

to result in significant amounts of diaddition [24]. *Thus, monoaddition would be expected for addition of organolithium compounds to 1,1-diphenylethylene, except when a large excess of DPE is used*; therefore, stoichiometric amounts of DPE should be utilized for most synthetic applications.

2.1.2
Kinetics and Mechanism

The kinetics of the addition reaction of alkyllithium compounds with 1,1-diphenylethylene as shown in Eq. (3) have been extensively investigated [19]:

$$\text{RLi} + \text{CH}_2=\overset{\underset{|}{C_6H_5}}{\underset{\underset{|}{C_6H_5}}{C}} \xrightarrow{k} \text{RCH}_2\overset{\underset{|}{C_6H_5}}{\underset{\underset{|}{C_6H_5}}{C}}\text{Li} \quad (3)$$

This addition reaction with DPE is of interest as a model for the initiation step in alkyllithium-initiated polymerization without the added complication of propagation. Another advantage is that the diphenylalkyllithium adduct exhibits strong UV-visible absorbances at $\lambda_{max}=428$ nm ($\varepsilon=1.6\times10^4$), $\lambda_{max}=438$ nm ($\varepsilon=2.5\times10^4$), and $\lambda_{max}=495$ nm ($\varepsilon=2.8\times10^4$) in benzene [23], diethyl ether [25], and THF [26], respectively, which are well separated from the absorptions due to the alkyllithium initiators. The kinetics exhibit a first order dependence on the DPE concentration, but fractional kinetic order dependencies on the alkyllithium concentrations have generally been observed as shown in Table 1. For example, in benzene a one-sixth order dependence on the n-butyllithium concentration has been reported [23], while a one-fourth order dependence has been observed for *tert*-butyllithium [28]. In most cases these fractional kinetic orders correspond to the reciprocals of the degrees of association (N) of the organolithium compound, which are also listed in Table 1 [3, 8, 19]. These results have most simply been rationalized by proposing that dissociation of the kinetically inactive organolithium aggregates to a kinetically active monomeric (unassociated) species is required prior to addition to DPE as shown in Scheme 2. However, a kinetic order of 0.65 for phenyllithium addition to DPE in THF has been interpreted in terms of an equilibrium between monomeric and dimeric phenyllithium species and the proposal that both species react with DPE [26].

It is prudent to note that Brown [31] has pointed out that the mechanism shown in Scheme 2 is not consistent with available experimental evidence regarding the kinetics and energetics of dissociation of alkyllithium aggregates. Thus, there is no direct experimental evidence which demonstrates the existence of species such as monomeric n-butyllithium as a distinct molecular species in hydrocarbon solvents [8]. In addition, ab initio calculations predict that the enthalpy of dissociation of tetramers into dimers is 144–152 kJ/mol

Table 1. Degrees of aggregation (N) of alkyllithiums and kinetic reaction orders (n) for addition of alkyllithiums (in excess) with 1,1-diphenylethylene

RLi	N^a	Solvent	n	$\Delta G^{\neq b}$	$\Delta H^{\neq b}$	$\Delta S^{\neq c}$	Ref
n-C_4H_9Li	6	Benzene	0.18	23.8	15.8	−26.4	23, 27
C_2H_5Li	6		∼0.11	24.2	17.3	−22.8	27
tert-C_4H_9Li	4		0.25	21.0	15.5	−18.0	28
$ArCH_2Li^d$	2^e	Toluene	0.5^f	−	−	−	29
n-C_4H_9Li	−	0.4% Et_2O/C_6H_6	0.50	20.7	9.4	−37.3	30
C_2H_5Li	−		0.25	21.4	9.4	−39.6	30
C_6H_5Li	−	4% Et_2O/C_6H_6	0.41	23.7	14.9	−29.1	30
$C_6H_5CH_2$Li	−	Et_2O	1.2	−	−	−	25
CH_3Li	4		0.21	−	−	−	25
CH_2=CHCH$_2$Li	2		1.3	−	−	−	25
C_6H_5Li	−		0.51	−	−	−	25
n-C_4H_9Li	4		0.3	−	−	−	25
n-C_4H_9Li	2.4-2.8^g	THF	0.5	16.1	7.9	−27.2	30
n-C_4H_9Li			∼0.4	−	−	−	26
CH_2=CHLi	4^h		0.34	−	−	−	26
$C_6H_5CH_2$Li	1		1.1	−	−	−	26
CH_3Li	4		0.27	−	−	−	26
CH_2=CHCH$_2$Li	2.1^g		∼1	−	−	−	26
C_6H_5Li	2		0.66	−	−	−	26

^{a} See [3, 8, 19].
^{b} kcal/mol.
^{c} cal/mol deg.
^{d} p-tert-Amylbenzyllithium.
^{e} Degree of aggregation assumed to be the same as benzyllithium in benzene.
^{f} Kinetics effected with excess DPE (0.24-5.9 molar excess) and in the presence of residual diethyl ether.
^{g} −108 °C.
^{h} Degree of aggregation of trans-1-propenyllithium.

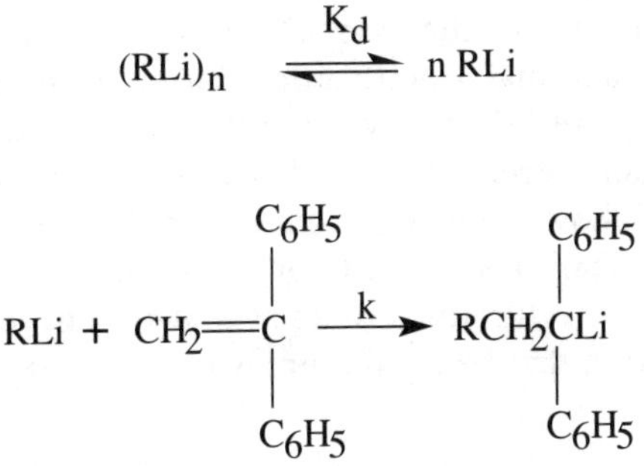

$$(RLi)_n \xrightleftharpoons{K_d} n\,RLi$$

$$RLi + CH_2{=}C{\begin{smallmatrix}C_6H_5\\ \\C_6H_5\end{smallmatrix}} \xrightarrow{k} RCH_2CLi{\begin{smallmatrix}C_6H_5\\ \\C_6H_5\end{smallmatrix}}$$

Scheme 2.

$$(RLi)_6 \rightleftharpoons (RLi)_4 + (RLi)_2$$
$$(RLi)_6 \rightleftharpoons (RLi)_5 + RLi$$
$$(RLi)_4 \rightleftharpoons 2(RLi)_2$$
$$(RLi)_2 \rightleftharpoons 2 RLi$$

Scheme 3.

while the enthalpy of dissociation to the unaggregated species is 420.4–514.6 kJ/mol [32, 33]. In contrast to these large predicted heats of dissociation, the observed activation parameters for addition of alkyllithiums to 1,1-DPE are much smaller as shown in Table 1. For example, the enthalpy of activation for addition of n-butyllithium to DPE in benzene is only 66.2 kJ/mol. This number appears to be too low to involve dissociation of the hexameric aggregate prior to addition to DPE.

An alternative interpretation of these fractional kinetic orders in alkyllithium concentration as proposed by Wakefield [19] is that the rate-determining step involves coordination of a DPE molecule to one face of the polyhedral organolithium aggregate. As suggested previously [8], incomplete or stepwise dissociation equilibria such as those shown in Scheme 3 would be expected to require less energy as predicted by theoretical calculations [32]. It is important to note that Brown and coworkers [34, 35] have reported that dissociation energies for tetramer-dimer equilibria are 46.1 kJ/mol and 100 kJ/mol for methyllithium in ether [34] and *tert*-butyllithium in cyclopentane [35], respectively.

A further problem that complicates the kinetics of these addition reactions is the fact that cross-association [3] of the diphenylalkyllithium species (dimeric in hydrocarbon solution) [36] with the alkyllithium (tetrameric or hexameric association) [3, 37–40] would be expected to change the average association state of the predominant form of the initiator as the reaction proceeds [3, 8, 40]. Thus, interpretable kinetic orders may only be available for the initial part of the kinetics. Fortunately, most kinetic studies have utilized a large excess of alkyllithium relative to DPE which not only minimizes the effect of this cross-association but effectively eliminates the problem of oligomerization of DPE (Scheme 1).

A further consequence of the association of alkyllithium compounds and the corresponding fractional kinetic orders for addition to DPE is the observation that the relative reactivities of alkyllithium compounds with 1,1-DPE in THF are concentration dependent [25]. Thus, structure-reactivity correlations do not necessarily follow from consideration of the expected acidities of the corresponding hydrocarbons [3]. At low concentrations, n-butyllithium is more reactive than benzyllithium, while at higher concentrations it was predicted by extrapolation that benzyllithium would be more reactive than n-butyllithium.

The reactivity of DPE relative to styrene can be determined by comparison of their rates of reaction with n-butyllithium in benzene. The initiation rate constant for styrene at 30.3 °C was reported to be 2.33×10^{-5} (l/mol)$^{1/6}$ s^{-1} [41], while the corresponding rate constant for addition to DPE was reported to be 4.17×10^{-5} (l/mol)$^{1/6}$ s^{-1} [23]. Thus, DPE is about twice as reactive as styrene with respect to addition of n-butyllithium.

Thus this model system for initiation, the reaction of alkyllithium initiators with DPE, is still quite complicated primarily because of the association of the organolithium chain ends and the cross-association of residual alkyllithium with the product, 1,1-diphenylalkyllithium. In spite of these complications, several interesting and important conclusions can be gleaned from these studies. In general there is an inverse relationship between the reactivity of a given organolithium compound and its degree of association, i.e., less associated species are more reactive. Thus, tetrameric *tert*-butyllithium is 200 times more reactive than hexameric n-butyllithium [28]. Furthermore, this generalization can be extended further by noting that addition of Lewis bases such as amines and ethers generally decreases the average degree of association of an alkyllithium or eliminates association completely [3, 8, 19, 42]. As shown in Table 1, the reactivity of alkyllithiums is enhanced in the presence of ethers as shown by lower enthalpies of activation and increases in the magnitudes of the exponents for the kinetic orders with respect to the alkyllithium concentration.

2.2
Stability and Structure of 1,1-Diphenylalkyllithiums

Insight into the chemistry and reactivity of 1,1-diphenylalkyllithiums can be obtained by examination of their stability and structure. A carbanion can be defined as the species formed by removal of a proton from a carbon acid. For the diphenylmethyl carbanion, the corresponding carbon acid would be diphenylmethane as shown in Eq. (4):

$$(C_6H_5)_2CH_2 \underset{}{\overset{K_{eq}}{\rightleftharpoons}} (C_6H_5)_2CH:^{\ominus} + H^{\oplus} \qquad (4)$$

Diphenylmethane is the conjugate acid of the diphenylmethyl carbanion, and the equilibrium acidity constants (K_a) have been measured both directly and indirectly in the gas phase and in solution [3]. The most extensive investigations of the effect of structure on acidity for carbon acids have been carried out in DMSO using a carbon indicator method to determine relative acidities and this scale was anchored with potentiometric measurements to provide an absolute scale of acidities [3, 43]. A summary of relevant pK_a values for various carbon acids is shown in Table 2. The data in Table 2 are especially relevant for considering the reactivity of 1,1-diphenylmethyl carbanionic species as initiators in anionic polymerization. In general, an appropriate initiator for a given monomer is an anionic species that has a reactivity (stability) similar

Table 2. Acidities of carbon acids[a] in DMSO at 25 °C [pK_a(DMSO)]

Carbon acid	pK_a (DMSO)	Reference
Fluorene	22.6	44
Triphenylmethane	30.6	44
Ethyl acetate	30–31[a]	45
Diphenylmethane	32.2	43
Dimethylsulfoxide	35.1	43
Toluene	43[b]	43
Propene	44[b]	43
Methane	56[b]	43

[a] A carbon acid has a carbon-hydrogen covalent bond which can dissociate to form a carbanion and a proton; a carbon acid is the conjugate acid of the corresponding carbanion (see [3], p 33).
[b] Estimated by extrapolation.

to the propagating carbanionic species [3, 7]. Thus, it would be expected that an effective initiator for a given monomer would have a pK_a value with the same or larger value than the conjugate acid of the propagating anionic species. Diphenylmethane has a pK_a (32.2) which is larger than, but similar to, the pK_a of ethyl acetate (30–31), which can be considered as the conjugate acid of the propagating ester enolate anion for polymerization of methyl methacrylate. Indeed, 1,1-diphenylalkylcarbanions are useful and effective initiators for anionic polymerization of alkyl methacrylates [3, 46–48]. Using the same logic, it would be predicted that 1,1-diphenylalkylcarbanions would not be effective initiators for the anionic polymerization of styrenes or dienes. The pK_a of diphenylmethane (32.2) is much smaller than the pK_a values of toluene (43) and propene (44), which correspond to the conjugate acids of benzyl and allyl carbanions. The benzyl and allyl carbanions can be considered as models for the propagating anions in anionic polymerization of styrenes and dienes, respectively. Contrary to these predictions, 1,1-diphenylmethylcarbanions are reactive and efficient initiators for the anionic polymerization of both styrene and diene monomers [3, 49, 50]. This result can be rationalized by noting that, in addition to the energetics of forming a less stable carbanion from a more stable carbanion, a π-bond is converted to a more stable σ-bond. This exothermic process must contribute significant stabilization to the transition state such that this initiation reaction is energetically favorable [51].

Insight into the structure of organolithium compounds can be deduced from their X-ray crystal structures, recognizing that the energetics in the solid state may perturb the structure compared to the structure in solution [3]. A model for 1,1-diphenylalkyllithium compounds has been provided by the 2-electron reduction of 1,1-diphenylethylene with lithium metal in diethyl ether as shown in Eq. (5) [52]:

$$2 \text{ Ph}_2\text{C}=\text{CH}_2 \xrightarrow{\text{Li, Et}_2\text{O}} \text{[dilithium 1,4-butane-diide dimer]} \quad (5)$$

In the observed structure, each lithium cation is coordinated to two ether molecules (not shown) and also π-coordinated both to the almost trigonal-planar trisubstituted anionic carbon centers and also to adjacent centers of the less twisted ($\omega=8°$) phenyl ring. The other phenyl ring is twisted by 37° out of the plane of the 1,4-butane-diide salt. The salt possesses a center of inversion, i.e., one lithium is above the plane of the primary structure and the other lithium is below this plane. The strong interaction between lithium and two of the aromatic carbons of the less twisted ring is shown by the bond distances in structure 1. An analogous interaction of lithium with both the C_2 carbon and one of the *ortho* carbons has been observed in the X-ray structure of a quinuclidene adduct of benzyllithium [53]. This structural information provides confirmation of the expectation that diphenylalkyl carbanions exhibit trigonal, planar structures (C_1-C_2, C_1-CH_2, C_1-C_2') and that the negative charge is delocalized into the phenyl rings.

NMR studies have provided insight into the hybridization and charge distribution for 1,1-diphenylalkyllithiums [54–57]. In general, proton chemical shifts are sensitive to the charge distribution on carbon; a conversion factor of 10 ppm/electron has been used to estimate charge distributions in carbanions [54, 57, 58]. Proton chemical shifts are generally shifted upfield by increased negative charge on the attached carbon. ^{1}H NMR studies of ring hydrogen chemical shifts for 1,1-diphenylhexyllithium indicate that there is more delocalization of charge in THF (0.87 electron) compared to cyclohexane (0.61 electron) [56]. An analogous study reported that the amount of charge delocalized into the rings corresponded to 0.50 electron in deuterated benzene and 0.78 electron in THF; similar charge distributions were also observed for 1,1-diphenyl-*n*-butyllithium [56]. For 1,1-diphenylmethyllithium it was estimated that the charge delocalized into the phenyl rings corresponded to 0.72

$\text{Li}-C_1 = 2.28$ Å

$\text{Li}-C_2 = 2.32$ Å

$\text{Li}-C_3 = 2.58$ Å

1

electron in THF [57]. The ^{13}C chemical shift for the α-carbon of diphenylmethyllithium in THF is reportedly shifted upfield by 43 ppm [54] or 39 ppm [59] relative to the methylene carbon in 1,1-diphenylmethane. In general, carbon chemical shifts are shifted upfield by an amount corresponding to 160–200 ppm/electron [54]. Upfield shifts are also observed upon rehybridization; an upfield shift of 110 ppm is observed for a hybridization change from sp^3 to sp^2 for carbon [54]. The J(^{13}C-H) coupling constant is a useful probe for hybridization as shown in Eq. (6):

$$J(^{13}\text{C-H}) = 5 \times (\% \text{ s}) \text{ [Hz]} \tag{6}$$

where % s represents the percentage s character of the carbon hybrid orbital participating in the C-H bond [60, 61]. An sp^3-hybridized carbon would be expected to exhibit a J(^{13}C-H)=125 Hz and an sp^2-hybridized carbon would be expected to exhibit J(^{13}C-H)=167 Hz. Thus, the α-carbon of 1,1-diphenylmethane exhibits J(^{13}C-H)=126 Hz while diphenylmethyllithium exhibits J(^{13}C-H)=142 Hz [54]. For comparison, the ^{13}C chemical shift of benzyllithium is shifted only 9 ppm relative to toluene [54] and the J(^{13}C-H) coupling constant is reported to be 131–134 Hz [55, 59, 62, 63]. These data for benzyllithium have been interpreted in terms of a pyramidal benzylic CH$_2$ group [59]. The combination of ^{13}C chemical shift and J(^{13}C-H) coupling constant for diphenylmethyllithium have been interpreted in terms of sp^2-hybridization for the α-carbon [54, 55]. Further evidence regarding the structure of 1,1-diphenylalkyllithiums can be deduced from UV-visible spectral studies. The UV spectral absorption maximum for 1,1-diphenylhexyllithium shifts from 410 nm in hexane to 496 nm in THF [55]. These results suggest that there is more interaction between lithium and the α-carbon in hexane, resulting in less charge delocalization into the phenyl rings compared to the charge distribution in THF. Of course, this interpretation has to be oversimplified, recognizing that 1,1-diphenylalkyllithiums are dimeric in hydrocarbon solution and presumably unassociated in THF [36].

In conclusion, the structure and charge distribution for a 1,1-diphenylalkyl carbanion and the corresponding dimers can be represented by structures 2, 3 and 4 below, in which the α-carbon is sp^2-hybridized (3). The delocalized system (2) including the C_1 carbons of the phenyl rings and the α-carbon would be expected to be sensibly planar; however, at any given time, one or both of the phenyl rings would be expected to be slightly twisted to minimize steric crowding with the alkyl group R and the other phenyl ring (2). A dynamic twisting of the two phenyl rings would be expected to occur. The charge distribution (2) will depend on the solvent and the presence of Lewis base additives. The charge would tend to be more concentrated on the α-carbon in hydrocarbon solution where the presence of the small, highly charged lithium cation would perturb the electron distribution. It should also be noted that 1,1-diphenylalkyllithiums are associated into dimers in hydrocarbon media (4) [36], but would be expected to be unaggregated in polar media. In polar media or in the presence of Lewis bases which solvate the lithium cation, interaction with the α-carbon would be reduced and more of the negative charge would be delocalized into the aromatic rings.

2.3
Initiators for Anionic Polymerization

As discussed in the previous section, a useful guide to choose appropriate initiators for anionic polymerization of a given type of monomer is that the initiator should have approximately the same structure and reactivity as the propagating anionic species, i.e., the pK_a of the conjugate acid of the propagating anion should correspond closely to the pK_a of the conjugate acid of the initiating species [3, 7, 64]. If the initiator is too reactive, side reactions between the initiator and monomer will often occur; if the initiator is not reactive enough, then the initiation reaction may be slow, inefficient, or not occur at all.

2.3.1
Alkyl Methacrylates

It would be expected that 1,1-diphenylalkylcarbanions would be useful initiators for the anionic polymerization of alkyl methacrylates, since models for the corresponding conjugate acids, i.e., diphenylmethane and ethyl acetate, have similar pK_a (DMSO) values of 32.2 [43] and 30–31 [45], respectively (see Table 2). Pioneering studies by Wiles and Bywater [47, 48] and Rempp and coworkers [46] demonstrated the usefulness of 1,1-diphenylalkylcarbanions as initiators for the controlled polymerization of alkyl methacrylates, especially methyl methacrylate (MMA). Wiles and Bywater [48] generated the 1,1-diphenylhexyllithium initiator by the quantitative and facile addition of butyllithium to 1,1-diphenylethylene (see Eq. 1). They used this initiator for methyl methacrylate polymerization in toluene at low temperatures. Unfortunately, the use of toluene results in complex product mixtures; broad and often multimodal molecular weight distributions, indicating a multiplicity of active propagating species, are typically obtained in toluene and include low molecular weight, soluble fractions [65–68]. However, in spite of these complications and in contrast to work with n-butyllithium as initiator, the kinetics of polymerization of MMA in toluene using the 1,1-diphenylhexyllithium initiator exhibited first-order dependence on both the monomer and the initiator concentrations [48]. Later work by Hatada and coworkers [68] has shown that the reaction of MMA with n-butyllithium in toluene at $-78\,°C$ produces approximately 51% of lithium methoxide by attack at the carbonyl carbon of methyl methacrylate. In contrast, only 17% lithium methoxide was formed when 1,1-diphenylhexyllithium was used in place of n-butyllithium. Rempp and coworkers [46] prepared a polymeric 1,1-diphenylalkylpotassium initiator in THF by addition of poly(styryl)potassium to DPE at $-78\,°C$ and demonstrated that this polymeric carbanion efficiently initiated MMA block copolymerization to form the corresponding block copolymers with narrow molecular weight distributions. Since these reports on the use of 1,1-diphenylalkylcarbanions as initiators for MMA polymerization, the adduct of butyllithium with 1,1-diphenylethylene has been the initiator of choice for controlled anionic polymerization of MMA at low temperatures [3, 69, 70]. Using this initiator system, controlled MMA polymerization can be effected to form PMMA with controlled M_n and narrow molecular weight distribution ($M_w/M_n=1.1$). Recent advances include the addition of lithium chloride [71–73], lithium *tert*-butoxide [74], lithium 2-(2-methoxyethoxy)ethoxide [75], or lithium butyldimethylsilanolate [76] to generate narrow molecular weight distributions and to improve the stability of active centers for anionic polymerization of both alkyl methacrylates and *tert*-butyl acrylate.

The adducts of either butyllithium with DPE (1,1-diphenylhexyllithium) or polymeric organolithiums with DPE are the preferred initiators for polymerization of a wide variety of alkyl methacrylates and *tert*-butyl acrylate [3]. For example, Stühn and coworkers [77] recently reported the synthesis of well-defined, narrow molecular weight distribution polymers from methyl, ethyl, n-propyl, n-butyl, n-pentyl, and n-hexyl methacrylates using the adduct of *sec*-butyllithium and DPE as initiator in THF at $-78\,°C$ in the presence of lithium chloride. This initiator was also used recently to copolymerize 2-trimethylsiloxyethyl methacrylate and butyl methacrylate using analogous reaction procedures [78].

2.3.2
Styrenes and Dienes

One of the most surprising and useful aspects of the organolithium chemistry of 1,1-diphenylethylene is that reactions with simple and polymeric organolithiums form the corresponding 1,1-diphenylalkyllithiums which are effective initiators for the anionic polymerization of styrene and diene monomers [64]. As described in Sect. 2.2, the pK_a (DMSO) value of diphenylmethane (32.2) is much lower than the estimated pK_as (DMSO) of toluene (43) and propene (44) (see Table 2) [43]. These pK_a differences correspond to an energy difference of ≥ 64.5 kJ/mol; thus, the diphenylmethyl carbanion is 64.5 kJ/mol more stable than either the benzyl carbanion or the allyl carbanion. Since these latter carbanions serve as models for the carbanionic propagating species which would be formed upon addition of a diphenylmethyl carbanion initiator to styrene (Eq. 7) and butadiene (Eq. 8), it would be expected that these initiation reactions would be energetically unfavorable:

$$(C_6H_5)_2CH^\ominus + CH_2{=}CH(C_6H_5) \xrightarrow{k_i} (C_6H_5)_2CHCH_2CH^\ominus(C_6H_5) \quad (7)$$

$$(C_6H_5)_2CH^\ominus + CH_2{=}CH{-}CH{=}CH_2 \xrightarrow{k_i}$$

$$(C_6H_5)_2CHCH_2{-}CH{=}CH{-}CH_2^\ominus$$
$$\updownarrow$$
$$(C_6H_5)_2CHCH_2{-}\overset{\ominus}{CH}{-}CH{=}CH_2 \quad (8)$$

As noted previously, the fallacy of this argument is that an accounting of the energetics of these initiation reactions must also include the energetics of converting a π-bond in the monomer to a more stable σ-bond [51] in the adduct, and this is the primary driving force for all vinyl polymerizations.

Before discussing the use of 1,1-diphenylalkylcarbanions as initiators for the anionic polymerization of styrenes and dienes, it is important to consider the characteristics of a useful anionic initiator. Several criteria have been proposed for judging whether an organolithium initiator is useful for preparation of well-defined polymers [79, 80]:

Criterion 1. The initiator should be soluble in hydrocarbon media. Hydrocarbon solubility provides the possibility to prepare polydienes with high 1,4-microstructure [3, 81].

Criterion 2. The initiator must efficiently initiate chain-growth polymerization such that all of the initiator is consumed before monomer consumption is complete. Initiator efficiency is essential to prepare polymers with controlled molecular weight. If the initiator is efficient, then the observed number average molecular weight will be in agreement with the value calculated based on the stoichiometry of the polymerization as shown in Eq. (9) [3, 82]:

$$M_n = \frac{\text{g of monomer polymerized}}{\text{moles of RLi initiator}} \quad (9)$$

Criterion 3. The rate of initiation should be competitive with or faster than the rate of propagation ($R_i \geq R_p$) [82]. As delineated by Flory [83, 84], it is possible to obtain a Poisson-like, narrow molecular weight distribution ($M_w/M_n \leq 1.1$) if all of the chains are initiated at approximately the same time and grow for approximately the same time (no termination and no irreversible chain transfer). Under these conditions the molecular weight distribution will obey the following relationship (Eq. 10):

$$X_W/X_n = 1 + \frac{X_n}{(X_n - 1)^2} \tag{10}$$

where X_w is the weight average degree of polymerization and X_n is the number average degree of polymerization. When the number average molecular weight is large relative to 1 ($X_n \gg 1$), Eq. (10) reduces to Eq. (11):

$$X_W/X_n = 1 + \frac{1}{X_n} \tag{11}$$

These three criteria provide a basis upon which to judge the usefulness of 1,1-diphenylethylene-derived organolithium initiators for polymerization of styrene and diene monomers in hydrocarbon solution.

The first publication on the use of 1,1-diphenylalkyllithiums as initiators for diene or styrene polymerization is the pioneering study of Morton and Fetters [49] on the preparation of poly(α-methylstyrene-*block*-isoprene-*block*-α-methylstyrene) using the dilithium dimer of 1,1-diphenylethylene, 1,4-dilithium-1,1,4,4-tetraphenylbutane (**6**) (see Scheme 4). The dilithium initiator

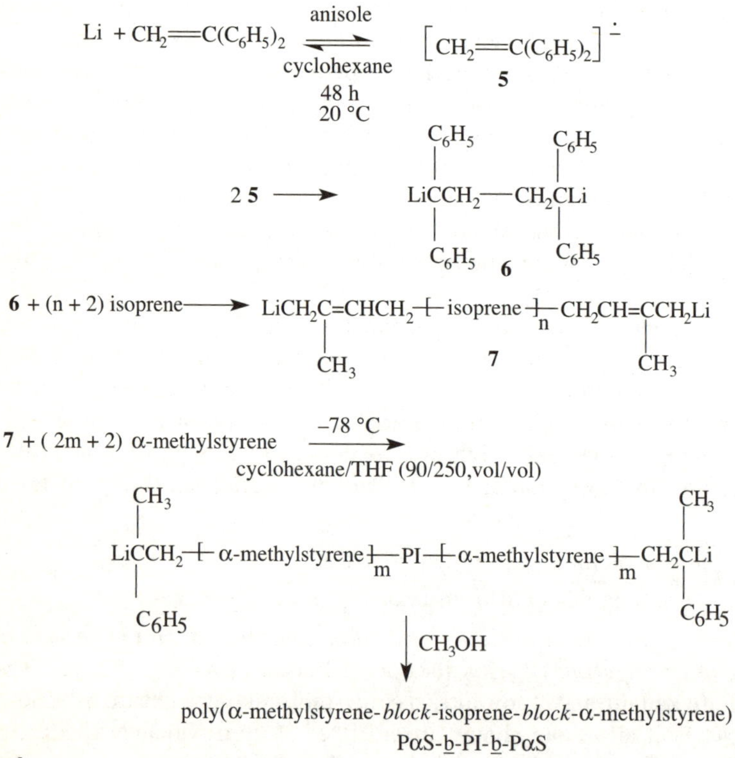

$$Li + CH_2=C(C_6H_5)_2 \xrightleftharpoons[\text{cyclohexane}]{\text{anisole}} \underset{\text{48 h}}{\underset{20\,°C}{}} [CH_2=C(C_6H_5)_2]^{\cdot -}$$
5

$$2\,\mathbf{5} \longrightarrow \underset{\underset{C_6H_5}{|}}{\overset{\overset{C_6H_5}{|}}{LiCCH_2}} - \underset{\underset{C_6H_5}{|}}{\overset{\overset{C_6H_5}{|}}{CH_2CLi}}$$
6

$$\mathbf{6} + (n+2)\,\text{isoprene} \longrightarrow \underset{\underset{CH_3}{|}}{LiCH_2C=CHCH_2} \!-\!\!\left[\text{isoprene}\right]_n\!\!-\!CH_2CH=\underset{\underset{CH_3}{|}}{CCH_2Li}$$
7

$$\mathbf{7} + (2m+2)\,\alpha\text{-methylstyrene} \xrightarrow[\text{cyclohexane/THF (90/250, vol/vol)}]{-78\,°C}$$

$$\underset{\underset{C_6H_5}{|}}{\overset{\overset{CH_3}{|}}{LiCCH_2}}\!-\!\!\left[\alpha\text{-methylstyrene}\right]_m\!\!-\!PI\!-\!\!\left[\alpha\text{-methylstyrene}\right]_m\!\!-\!\underset{\underset{C_6H_5}{|}}{\overset{\overset{CH_3}{|}}{CH_2CLi}}$$

$$\Big\downarrow CH_3OH$$

poly(α-methylstyrene-*block*-isoprene-*block*-α-methylstyrene)
PαS-b-PI-b-PαS

Scheme 4.

was formed by electron transfer [6, 9] from lithium metal to 1,1-diphenylethylene to form the corresponding radical anion, **5**, which rapidly dimerizes to form **6** [85–87]. The initiator was formed in a solvent mixture of cyclohexane and anisole (15 vol. %). With respect to the criteria delineated above to evaluate the 1,1-diphenylethylene-based initiators:

Criterion 1. The dilithium initiator, **6**, was soluble (0.01–0.08 mol/l) in a mixture of cyclohexane/anisole (85/15, v/v), but like most dilithium initiators [88, 89] precipitated from solution when added to the polymerization solvent, cyclohexane. Therefore, the dilithium initiator was chain extended with approximately 30 units of isoprene to generate an isoprenyldilithium oligomer which was soluble when added to cyclohexane. Since the final anisole concentration was less than 1 vol. %, a high (>85%) 1,4-polyisoprene center block was obtained (65–70% *cis*-1,4, 20–25% *trans*-1,4 and the remainder was 3,4-enchainment).

Criterion 2. The number average molecular weights of polyisoprenes and poly(α-methylstyrene-*block*-isoprene-*block*-α-methylstyrene) block copolymers prepared with the initiator, **6**, were all within 4% of the number average molecular weights calculated based on the stoichiometry of the reaction (see Eq. 9; note this equation should be modified to Eq. 12 if the moles of dilithium initiator is used in the denominator of Eq. (9)]:

$$M_n = \frac{\text{g of monomer polymerized}}{0.5 \text{ moles of dilithium initiator}} \qquad (12)$$

Criterion 3. All of the polyisoprenes and poly(α-methylstyrene-*block*-isoprene-*block*-α-methylstyrene) block copolymers prepared using the isoprene chain-extended derivative of **6** exhibited monomodal molecular weight distributions as observed by SEC. In addition, the triblock copolymers prepared using this initiator exhibited polydispersities (M_w/M_n) equal to 1.07, based on light scattering and osmometric molecular weight determinations.

In terms of these criteria, the initiator 1,4-dilithium-1,1,4,4-tetraphenylbutane (**6**) is a useful anionic initiator for the polymerization of isoprene monomer in hydrocarbon media. In fact, Yuki and Okamoto [90] reported that 1,1-diphenylhexyllithium reacts quantitatively with one mole of isoprene in benzene at room temperature. Apparently the steric effects of the diphenylalkyl group reduce the rate of oligomerization. Guzman and coworkers [91] investigated the kinetics of the anionic polymerization of isoprene in *n*-hexane and benzene initiated by 1,1-diphenyl-*n*-hexyllithium. They reported that the initiation reaction started immediately upon mixing and that it was complete in 3–10 min. Thus, these initiators are amazingly reactive for isoprene polymerization. In fact, it may be deduced that this initiator is more reactive than *n*-butyllithium in spite of the fact that the basic butyl carbanion would be expected to be much more reactive. Thus, at 30 °C, Hsieh [92] has reported that approximately 50% of *n*-butyllithium (2.6×10^{-3} mol/l) remains when all of the isoprene (1.3 mol/l) is completely polymerized in cyclohexane. Since Fetters and Young [36] have reported that 1,1-diphenylalkyllithiums are associated into dimers in hydrocarbon solution while *n*-butyllithium is associated into hexamers [3, 19, 38, 93], this reactivity pattern is in accord with the general trend that the reactivity of alkyllithiums increases as the degree of association decreases [3, 94].

One would expect that these 1,1-diphenylalkyllithium initiators would also be effective for butadiene polymerization. However, it is not clear from these results that 1,1-diphenylalkyllithium initiators are satisfactory for the polymerization of styrene monomers. These subjects will be discussed in Sect. 4.

3
Reactions of Polymeric Organolithium Compounds with 1,1-Diphenylethylene

Many interesting and important synthetic applications of 1,1-diphenylethylene and its derivatives in polymer chemistry are based on the addition reactions of polymeric organolithium compounds with 1,1-diphenylethylenes. Therefore, it is important to understand the scope and limitations of this chemistry. In contrast to the factors discussed with respect to the ability of 1,1-diphenylalkylcarbanions to initiate polymerization of styrenes and dienes, the additions of poly(styryl)lithium and poly(dienyl)lithium to 1,1-diphenylethylene should be very favorable reactions since it can be estimated that the corresponding 1,1-diphenylalkyllithium is approximately 64.5 kJ/mol more stable than allylic and benzylic carbanions as discussed in Sect. 2.2 (see Table 2). Furthermore, the exothermicity of this addition reaction is also enhanced by the conversion of a π-bond to a more stable σ-bond [51]. However, the rate of an addition reaction cannot be deduced from thermodynamic (equilibrium) data; an accessible kinetic pathway must also exist [3]. In the following sections, the importance of these kinetic considerations will be apparent.

3.1
Kinetics

The kinetics of the addition of polymeric organolithium compounds to 1,1-diphenylethylene (Eq. 13) would be expected to be analogous to the kinetics of the corresponding reaction with simple alkyllithiums (Eq. 3):

$$PLi + CH_2=C(C_6H_5)_2 \xrightarrow{k} PCH_2CLi(C_6H_5)_2 \qquad (13)$$

$$\mathbf{8}$$

Thus, monoaddition of one 1,1-diphenylethylene to the polymeric chain end would be expected, except when a large excess of DPE is employed [20–23].

The rate and efficiency of these addition reactions can be readily monitored by ultraviolet-visible spectroscopy, because the adduct 1,1-diphenylalkyl-lithium (8) absorbs at λ_{max}=440 nm in benzene and cyclohexane [23, 95], which is well separated from the absorption peak maxima for poly(styryl)-lithium (λ_{max}=334 nm in benzene [96] and 328 nm in cyclohexane [97]) and for poly(butadienyl)lithium (λ_{max}=272 nm in cyclohexane [97]). Because poly(styryl)lithium, poly(dienyl)lithium, and polymeric 1,1-diphenylalkyl-lithium [36] chain ends are associated in hydrocarbon solution to at least dimers [3], the kinetics would be expected to be complicated by cross-aggregation of the polymeric organolithium chain ends with the chain ends of the polymeric 1,1-diphenylalkyllithium adducts (8) as shown schematically in Eq. (14):

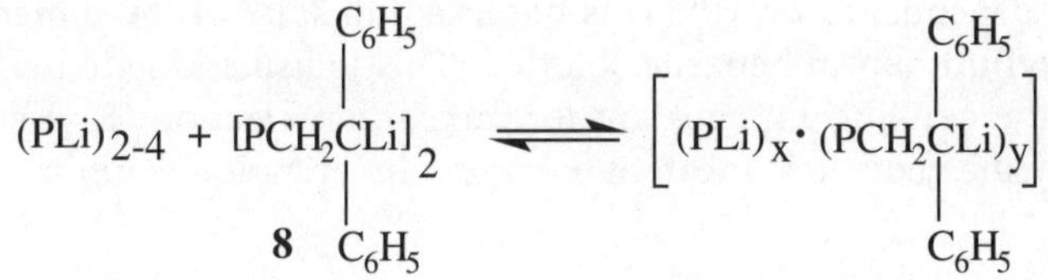

$$(\text{PLi})_{2\text{-}4} + [\text{PCH}_2\overset{\underset{C_6H_5}{|}}{\underset{\underset{C_6H_5}{|}}{C}}\text{Li}]_2 \rightleftharpoons \left[(\text{PLi})_x \cdot (\text{PCH}_2\overset{\underset{C_6H_5}{|}}{\underset{\underset{C_6H_5}{|}}{C}}\text{Li})_y\right] \quad (14)$$

$$8$$

3.2
Poly(styryl)lithium

Laita and Szwarc [98] have investigated the kinetics of the addition of poly(styryl)lithium to excess 1,1-diphenylethylene (4–1153 molar excess) in benzene. It was assumed that only monoaddition occurred, although previous work suggests that oligomerization may occur at higher ratios of DPE to organolithium [22, 23] as discussed in Sect. 2.1.1. Fortunately, even if oligomerization did occur, it would not be expected to affect the kinetic results significantly. The observed kinetics were dominated by the presence of equilibria between homodimers of poly(styryl)lithium (K_1), homodimers of the polymeric 1,1-diphenylalkyllithium (K_2), and the mixed dimers (K_{12}) as shown in Scheme 5. Under these conditions, the addition reaction was first-order in DPE, proportional to the concentration of unconverted PSLi and the observed

$$(\text{PSLi})_2 \;\overset{K_1}{\rightleftharpoons}\; 2\text{PSLi}$$

$$(\text{PS-DLi})_2 \;\overset{K_2}{\rightleftharpoons}\; 2\text{PS-DLi}$$

$$(\text{PSLi, PS-DLi}) \;\overset{K_{12}}{\rightleftharpoons}\; \text{PSLi} + \text{PS-DLi}$$

Scheme 5.

$$(PSLi)_2 \xrightleftharpoons{K_1} 2PSLi$$

$$PSLi + styrene \xrightarrow{k_{11}} PS\text{-}SLi$$

Scheme 6.

rate constant was inversely proportional to the initial concentration of PSLi ($[PSLi]_o$) as shown in Eq. (15):

$$\frac{-d[PSLi]}{dt} = \frac{k_{sd}K_1^{1/2}[PSLi][DPE]}{[PSLi]_0^{1/2}} \tag{15}$$

Thus, a first-order dependence on [PSLi] is observed, in spite of the dimeric nature of poly(styryl)lithium in benzene solution. This unusual kinetic behavior requires that the equilibrium constant for mixed dimerization (K_{12}) can be approximated by the geometric mean of the homodimerization equilibrium constants (Eq. 16):

$$K_{12} = (K_1 K_2)^{1/2} \tag{16}$$

Also implicit in the derivation of this kinetic equation is the assumption that PSLi present in the self-associated and cross-associated states is unreactive toward the addition to DPE [99].

The kinetic results of Laita and Szwarc [98] provide insight into the reactivity of 1,1-diphenylethylene. The rate constant for addition of DPE to monomeric poly(styryl)lithium ($k_{sd}K_1^{1/2}=2.1\times10^{-2}\,M^{-1/2}s^{-1}$) can be compared to the rate constant for styrene propagation by poly(styryl)lithium ($k_{11}K_1^{1/2}=9.4\times10^{-3}\,M^{-1/2}s^{-1}$) [97] as described in Scheme 6. This indicates that DPE is twice as reactive as styrene monomer with respect to addition to PSLi. This same reactivity difference was observed for addition of n-butyllithium to 1,1-diphenylethylene compared to styrene [23, 41].

3.2.1
Substituent Effects

Busson and van Beylen [95] investigated the kinetics of addition of excess poly(styryl)lithium to ring-substituted 1,1-diphenylethylenes. The expected one-half order kinetic dependence on [PSLi] was observed (Eq. 17):

$$\frac{-d[DPE]}{dt} = k_{obs}[PSLi]_{tot}^{1/2}[DPE] \qquad \left(k_{obs} = k_{sd}K_s^{1/2}\right) \tag{17}$$

This result is consistent with the fact that poly(styryl)lithium exists predominantly as dimers in benzene solution [3] and the proposal that the reactive species for addition to DPE is unassociated PSLi which is present in low concentrations in equilibrium with the dimers (see Sect. 2.1.2). The general validity of this interpretation, i.e., the relationship between degree of association and kinetic order, however, is a subject of current controversy [99–103].

The results of Busson and van Beylen [95] were analyzed in terms of the Hammett equation (Eq. 18) [104]:

$$\text{Log}\frac{k_x}{k_o} = \rho\sigma \tag{18}$$

where k_x is the rate of addition of PSLi to a 1,1-diphenylethylene substituted on the aromatic ring with a substituent x, k_o is the rate of addition of PSLi to unsubstituted 1,1-diphenylethylene, ρ (rho) represents the sensitivity of the reaction to the effect of changing the substituents, and σ (sigma) is a substituent constant. The substituent constants in the Hammett relationship are defined in Eq. (19):

$$\text{Log}\frac{K_x}{K_o} = \rho\sigma = \sigma \qquad \rho = 1 \tag{19}$$

in which the logarithm of the ratio of the equilibrium constants for ionization of substituted benzoic acid (K_x) relative to the equilibrium constant for ionization of unsubstituted benzoic acid (K_o) is used to define the substituent constants, σ, by defining ρ as having a value of unity for the ionization of benzoic acids (Eq. 19). Thus, the effect of substituents on a given reaction is always compared with the effects of substituents on the ionization of benzoic acid as a model, for which ρ is equal to one. Since the ionization of benzoic acids would be enhanced by electron-withdrawing substituents, reactions (or equilibria) which are favored by electron-withdrawing substituents would also be expected to exhibit a positive value of ρ and the magnitude of ρ would indicate whether this reaction (or equilibrium) is more or less sensitive to the effect of substituents than the ionization of benzoic acids. Conversely, if a reaction or equilibrium is favored by electron-donating groups, a negative value of ρ would be expected and the magnitude of ρ would indicate how sensitive the reaction or equilibrium is to the effect of substituents.

The ρ value for addition of poly(styryl)lithium with substituted 1,1-diphenylethylenes varied from +1.7 to +1.9 in benzene and cyclohexane [95]. The average value of 1.8 indicates that this reaction is twice as sensitive to substituents as the ionization of benzoic acids (ρ=1). This enhanced sensitivity would be expected, since the developing negative charge in the transition state can be conjugated directly with the substituents, whereas no direct conjugation of the negative charge with substituents on the aromatic ring can occur in the benzoic acid system. Perhaps the surprising aspect of this result is that the ρ value is not larger. Rho (ρ) values of +2.2 and +2.4 were reported for addition of poly(styryl)potassium and poly(styryl)cesium, respectively, with substituted 1,1-diphenylethylenes in benzene [95]. The ρ value varied from 2.8–3.5 for the poly(styryl)potassium addition to substituted DPEs in THF. For comparison, a positive ρ value (ρ=+1) was found for the initiation reaction of substituted styrenes using *n*-butyllithium as initiator in benzene [105]. A

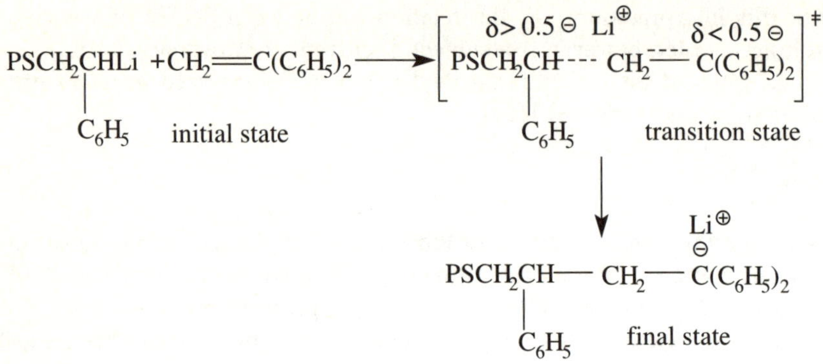

$$\text{PSCH}_2\text{CHLi} + \text{CH}_2\!=\!\text{C}(\text{C}_6\text{H}_5)_2 \longrightarrow \left[\underset{\text{transition state}}{\underset{\underset{\text{C}_6\text{H}_5}{|}}{\text{PSCH}_2\text{CH}}\cdots\overset{\delta>0.5\ominus\quad \text{Li}^\oplus}{\text{CH}_2}\!\!=\!\!\overset{\delta<0.5\ominus}{\text{C}(\text{C}_6\text{H}_5)_2}} \right]^{\ddagger}$$

initial state

$$\downarrow$$

$$\underset{\text{final state}}{\underset{\underset{\text{C}_6\text{H}_5}{|}}{\text{PSCH}_2\text{CH}}\text{---}\text{CH}_2\text{---}\overset{\text{Li}^\oplus\;\ominus}{\text{C}(\text{C}_6\text{H}_5)_2}}$$

Scheme 7.

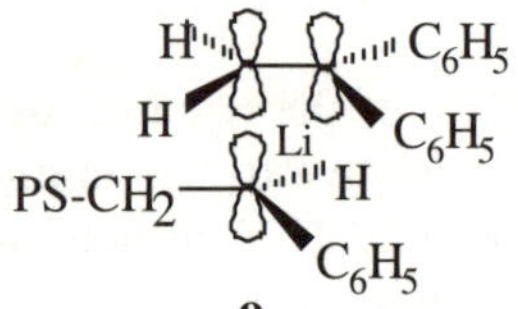

9

larger, positive ρ value was reported for anionic polymerization of substituted styrenes with sodium counterion in THF (ρ=+5.0) [106].

In interpreting these results, several factors must be considered. One of the most important considerations is the structure of the transition state (see Scheme 7). Based on the facts that a more stable 1,1-diphenylalkyl carbanion is formed from a benzyl carbanion and also a π-bond is converted into a σ-bond, the addition of PSLi to DPE should be an exothermic reaction. According to the Hammond Postulate [107], the structure of the transition state for an exothermic reaction should resemble the reactants. Thus, the amount of σ-bond making and π-bond breaking should have progressed less than 50%, i.e., more like reactants than products. An orbital description of the reaction is shown in structure **9**.

Consideration of the effect of counterion on the ρ value for addition of poly(styryl)carbanions to DPE is also instructive. The ρ value increases as the size of the counterion increases, i.e., as the electrostatic interaction of the counterion with the negative charge decreases. This effect would be expected for any transition state structure, since the small lithium cation would be expected to minimize delocalization into the phenyl rings (see Sect. 2.2 of this review and Chap. 1 of [3]). Thus, with the cesium counterion, for any degree of charge transfer onto the 1,1-diphenylcarbon in the transition state, this charge would be more delocalized than for the corresponding lithium system. It would also be predicted that the ρ value for addition of butyllithium to styrene would be less than that of the PSLi/DPE system ($\rho=+1$ vs 1.8 for the DPE system).

Thus, because this reaction would be even more exothermic, it would be predicted based on the Hammond Postulate [107] that there would be less charge transfer onto the benzylic carbon in the transition state. The fact that ρ values in THF are much larger than in hydrocarbon solution can also be explained by the effect of solvent on the counterion; solvation of the counterion would decrease the interaction of the counterion with the carbanion which would result in more delocalization of charge. Or in terms of the Winstein spectrum of ionic species [3, 108], solvation of the counterion and an increase in the dielectric constant of the medium would shift the structure towards a more ionic species.

In conclusion, the effect of substituents on the addition of PSLi to DPE is consistent with an exothermic transition state in which less than half of the charge has been transferred to the 1,1-diphenylalkylcarbanion, and the σ-bond-making and π-bond-breaking have also progressed to less than 50%.

3.2.2
Synthetic Utility

Lee [109] has investigated the rates of addition of polymeric organolithium compounds to 1,1-diphenylethylene from the point of view of synthetic utility. The addition of poly(styryl)lithium with stoichiometric amounts of DPE was complete at room temperature in 6 h in benzene and in 8 h in cyclohexane. This addition reaction was found to be independent of molecular weight in the range of 1500–8500. In order for this reaction to be synthetically useful for preparation of well-defined polymers, it is important to establish that this end-capping reaction is quantitative and irreversible for stoichiometric amounts of reactants. This question was investigated by using the UV-visible absorbances and absorption coefficients for poly(styryl)lithium (λ_{max}=334 nm; $\varepsilon=1.30\times10^4$ l/mol cm) [96] and the corresponding polymeric 1,1-diphenylalkyllithium (λ_{max}=440 nm; $\varepsilon=1.60\times10^4$ l/mol cm) [23, 95] formed by addition of a stoichiometric amount of 1,1-diphenylethylene to calculate the efficiency of the end-capping reaction. The results shown in Table 3 are consistent with almost quantitative end-capping. It was also established that the 1,1-diphenylalkyllithium was stable in benzene for one week at room temperature as judged by the constancy of the UV-visible absorbance.

Further evidence for the quantitative addition of poly(styryl)lithium with a stoichiometric amount of 1,1-diphenylethylene can be deduced by ^{1}H NMR analysis of the adduct formed after methanol termination. A characteristic peak (multiplet) at δ 3.5 ppm is observed by ^{1}H NMR for the terminal methine hydrogen at the chain end in the PS-DPE and no corresponding peak is observed in the base polystyrene. Integration of the area of this peak relative to the area of the resonances corresponding to the methyl protons of the *sec*-butyllithium initiator fragment at δ 0.5–0.78 ppm gave a value of 1:5.9, in close

Table 3. UV-visible absorption analysis of the efficiency of the end-capping reaction of poly(styryl)lithiums with 1,1-diphenylethylene in benzene at 25 °C

M_n of PSLi (g/mol)	Analysis of end-capping reaction A_{334} (PSLi)[a]	A_{440} (PS-DPELi)[b]	[PSDPELi]/[PSLi][c]
4,500	1.430	1.730	0.983
10,000	1.425	1.759	1.00
10,000	1.421	1.696	0.970

[a] Absorbance of poly(styryl)lithium. The extinction coefficient of poly(styryl)lithium at 334 nm is $\varepsilon = 1.30 \times 10^4$ l/mol cm [96].
[b] Absorbance of 1,1-diphenylalkyllithium. The extinction coefficient of 1,1-diphenylalkyllithium at 440 nm is $\varepsilon = 1.60 \times 10^4$ l/mol cm) [95].
[c] [PSDPELi]/[PSLi]=[A(PS-DPELi)/ε(PS-DPELi)] /[A(PSLi)/ε(PSLi)].

agreement with the expected value of 6.0 [109]. Thus, all of this spectroscopic evidence is consistent with the quantitative addition of stoichiometric amounts of DPE to poly(styryl)lithium in hydrocarbon solvent at room temperature after a period of approximately 8 h.

3.3
Poly(dienyl)lithiums

The kinetics of the addition of poly(2,4-hexadienyl)lithium to 1,1-diphenylethylene has been investigated by two groups. Fetters et al. [99] observed that the rate of addition exhibited a first-order dependence on the concentration of poly(2,4-hexadienyl)lithium, but the apparent first-order rate constant for addition was inversely proportional to the initial concentration of poly(2,4-hexadienyl)lithium chain ends. However, in contrast to the dimeric degree of association of poly(styryl)lithium, it was determined by concentrated solution viscometry studies that 35–40% of the poly(2,4-hexadienyl)lithium chain ends were not associated. Thus, it was concluded therein and elsewhere [100, 101] that there is no necessary relationship between the degree of association of the chain ends and the fractional kinetic orders observed.

Wang and Szwarc [110] reexamined the kinetics of the addition of poly(2,4-hexadienyl)lithium to DPE. They reported that the kinetic plots were curved, implying that, in contrast to the analogous poly(styryl)lithium addition, the geometric mean relationship for the dimerization constants was not fulfilled, i.e., $K_{12} \neq (K_1 K_2)^{1/2}$ (see Eq. 16). Using only the data from the initial, linear portion of the rate plot, the pseudo-first-order rate constants were found to be proportional to the initial chain end concentration raised to the power 0.43 rather than 1.0 as observed by Fetters and coworkers [99]. Concentrated solution viscosity measurements were interpreted in terms of dimeric association of the chain ends [110].

Lee [109] has also investigated the rates of addition of poly(dienyl)lithiums to 1,1-diphenylethylene from the point of view of synthetic utility. The rates of

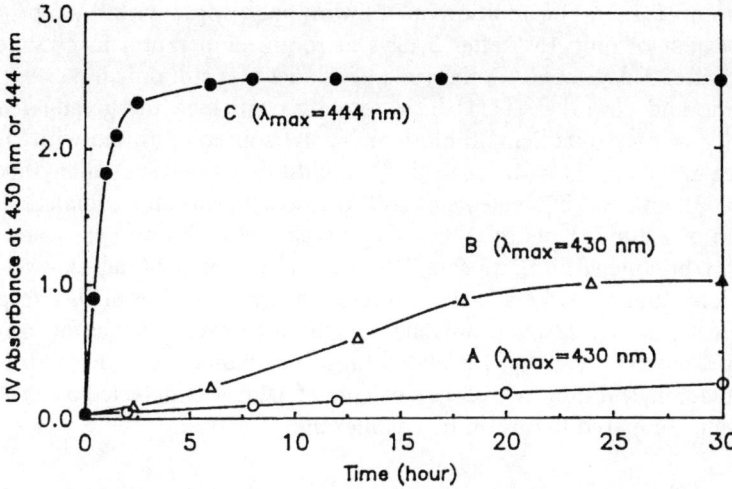

Fig. 1. Reaction rate plots for the addition of poly(isoprenyl)lithium (PILi) (M_n=1500 g/mol) to 1,1-diphenylethylene in cyclohexane with [PILi]=1.5×10^{-3} mol/l: (A) at 25 °C without THF; (B) at 50 °C without THF; (C) at 25 °C with THF ([THF]/[PILi]=40)

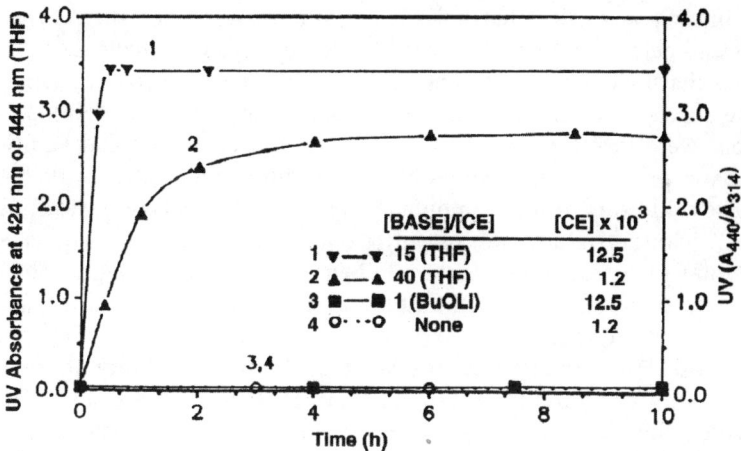

Fig. 2. Reaction rate plots for the addition of poly(butadienyl)lithium (PBDLi) (M_n=1400 g/mol) to 1,1-diphenylethylene in cyclohexane at 25 °C with [PBDLi]=1.1×10^{-3} mol/l

addition of poly(isoprenyl)lithium (M_n=1500) and poly(butadienyl)lithium (M_n=1400) with stoichiometric amounts of 1,1-diphenylethyene in cyclohexane are shown in Figs. 1 and 2, respectively. The reactivity of poly(dienyl)-lithiums with 1,1-diphenylethylene is unexpectedly slow compared to the analogous relatively rapid end-capping of poly(styryl)lithium [109]. After 3 days at 25 °C, the extent of end-capping with DPE was only 9% for poly(butadienyl)lithium (Fig. 2). The addition of DPE with poly(isoprenyl)lithium proceeds to the extent of only 15% after 3 days at room temperature in cyclohexane (Fig. 1); after 4 days at 50 °C, the yield of adduct was still only 60%.

Jerome and coworkers [111] have recently confirmed the greatly reduced reactivity of poly(butadienyl)lithium and poly(isoprenyl)lithium with DPE in cyclohexane using ^{7}Li NMR analysis. The addition of poly(butadienyl)lithium (M_n=5000) with two equivalents of DPE required 61 days for completion. The addition of 2 equivalents of DPE with poly(isoprenyl)lithium was faster than with poly(butadienyl)lithium; complete reaction required 14 days in cyclohexane. Jerome and coworkers [111] also reported that addition of DPE to poly(dienyl)lithiums was faster in toluene compared to cyclohexane; for example, the appearance of the diphenylalkyllithium resonance from the addition of poly(butadienyl)lithium with 5 equivalents of DPE was detected after 30 min in toluene compared to 60 min in cyclohexane.

3.3.1
Synthetic Utility

Thus, it is obvious that the addition of DPE to poly(dienyl)lithiums is not a synthetically useful reaction in hydrocarbon solution, even at elevated temperatures. Since it is known that poly(dienyl)lithiums are aggregated to at least dimers and perhaps tetramers or even higher aggregates in hydrocarbon solution and that less aggregated organolithiums are more reactive in general [3, 19], the effects of addition of the Lewis base, tetrahydrofuran (THF), on these reactions were investigated by Lee [109]. It has been shown that addition of Lewis bases promotes dissociation of organolithium aggregates [3, 19, 36, 112, 113]. Thus, addition of small amounts of THF ([THF]/[PLi]=40) was sufficient to effect quantitative addition of poly(isoprenyl)lithium to one equivalent of DPE within 4 h at room temperature in benzene as shown in Fig. 1. The addition of DPE to poly(butadienyl)lithium was complete within 1 h at 25 °C in the presence of 40 equivalents of THF as shown in Fig. 2. Only stoichiometric amounts of THF were utilized because of the known instability of polymeric organolithium compounds in THF [3, 111, 114]. In addition, the THF was added after the polymerization of the diene monomers was completed to preserve the high 1,4-diene microstructure which is unique to lithium counterion in hydrocarbon solution [3].

Similar observations have been reported by Jerome and coworkers [111]. These workers utilized 10 vol. % of either THF or diethyl ether. Chain end instability was observed as expected for the organolithium chain ends with this amount of THF at room temperature. However, it was observed that the chain ends were stable in the presence of a 1/9 diethyl ether/cyclohexane (v/v) mixture. A retarding effect of lithium alkoxides was also reported, in accord with the previous reports of Lochmann and coworkers [115].

It is interesting to speculate on the reasons for the relative unreactivity of poly(dienyl)lithium with DPE in hydrocarbon solvents. Since this relative unreactivity is eliminated in the presence of the Lewis base, THF (see also the discussion of copolymerization in the following section), the unreactivity in hydrocarbon solution does not reflect an intrinsic unreactivity of poly(dienyl)lithium chain ends with DPE. *It is proposed that this relative unreactivity is a consequence of the stronger association of poly(dienyl)lithium chain ends compared to poly(styryl)lithium chain ends which results in a lower extent of dissociation into reactive unassociated chain ends.* As discussed previously, poly(dienyl)lithium chain ends are associated into at least dimers and higher aggregates [3]. The effects of the addition of THF on the degree of association also indicates that the energetics of association are higher also, since much more THF is required to dissociate the poly(dienyl)lithium chain ends compared to poly(styryl)lithium chain ends [3, 8, 113]. In support of this conclusion, it should be noted that the crossover reaction rate of poly(2,4-hexadienyl)lithium to 1,1-DPE was found to be very fast [110]; the rate constant was one-half of the corresponding rate constant for addition of poly(styryl)lithium to DPE [98]. The association state of poly(2,4-hexadienyl)lithium has been reported to be 1.6 by Fetters and coworkers [99] and 2.0 by Wang and Szwarc [110]. Thus, the higher reactivity of poly(2,4-hexadienyl)lithium (PHDLi) compared to poly(butadienyl)lithium or poly(isoprenyl)lithium for addition to 1,1-DPE can be explained in terms of a lower degree of aggregation of PHDLi and a corresponding larger extent of dissociation to more reactive, unassociated species.

4
Polymeric 1,1-Diphenylalkyllithium Initiators for Block Copolymer Synthesis

Many of the applications of DPE in anionic polymerization chemistry require the adduct of a polymeric organolithium compound and DPE, e.g., a polymeric 1,1-diphenylalkyllithium (**10**, Scheme 8), to reinitiate polymerization of another monomer (M_b in Scheme 8). As discussed in Sect. 2.3.1, initiation of a methacrylate monomer which forms a much more stable ester enolate anion is favorable and rapid. A wide variety of amphiphilic block copolymers have been prepared by first polymerization of a styrene or diene monomer using an alkyllithium initiator, followed by end-capping with DPE and then block copolymerization of an alkyl methacrylate or acrylate [3]. For example, poly(butadiene-*block*-methyl methacrylate) and poly(isoprene-*block*-methyl methacrylate) block copolymers were prepared by *sec*-butyllithium-initiated polymerization of the diene in cyclohexane, followed by addition of DPE in the presence of 10 vol. % diethyl ether at room temperature and then addition of methyl methacrylate and THF (58 vol. %) at –78 °C [111]. Although the resulting butadiene block copolymer exhibited a somewhat broad molecular weight distribution (M_w/M_n=1.15), the isoprene block copolymers exhibited narrow molecular weight distributions (M_w/M_n=1.02, 1.04) under the same conditions.

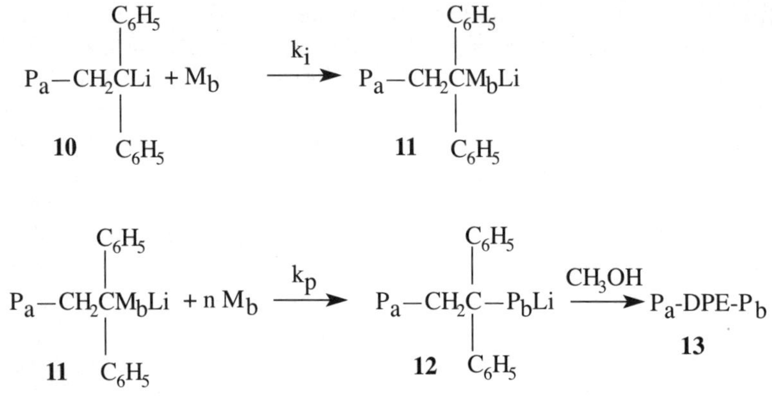

$$P_a-\underset{\underset{C_6H_5}{|}}{\overset{\overset{C_6H_5}{|}}{C}}H_2CLi + M_b \xrightarrow{k_i} P_a-\underset{\underset{C_6H_5}{|}}{\overset{\overset{C_6H_5}{|}}{C}}H_2CM_bLi$$
$$\quad\quad\quad 10 \quad\quad\quad\quad\quad\quad\quad\quad\quad\quad 11$$

$$P_a-\underset{\underset{C_6H_5}{|}}{\overset{\overset{C_6H_5}{|}}{C}}H_2CM_bLi + n\,M_b \xrightarrow{k_p} P_a-\underset{\underset{C_6H_5}{|}}{\overset{\overset{C_6H_5}{|}}{C}}H_2C-P_bLi \xrightarrow{CH_3OH} P_a\text{-DPE-}P_b$$
$$\quad\quad 11 \quad\quad\quad\quad\quad\quad\quad\quad\quad\quad 12 \quad\quad\quad\quad\quad\quad 13$$

Scheme 8.

A series of poly(styrene-*block*-n-butyl methacrylate) block copolymers have been prepared by *sec*-butyllithium initiated polymerization of styrene in THF at –78 °C, end-capping with DPE and then addition of *n*-butyl methacrylate [116]. Similarly, well-defined, smectic liquid-crystalline block copolymers were prepared by sequential polymerization of styrene, end-capping with DPE and block copolymerization of methacrylate monomers containing (S)-2-methylbutyl 4-(4-hydroxyphenylcarbonyloxy)biphenyl-4'-carboxylate mesogens at –78 °C in THF [117]. Using similar procedures, side-chain liquid crystalline polystyrene block copolymers have also been prepared from 6-[4-(4'-cyanophenyl)phenoxy]hexyl methacrylate [118]. Thermoplastic elastomeric poly(methyl methacrylate-*block*-butadiene-*block*-methyl methacrylate) triblocks have been prepared using the dilithium initiator formed from 2 moles of *sec*-butyllithium and *m*-diisopropenylbenzene in cyclohexane (5 vol. % diethyl ether) to polymerize butadiene, followed by end-capping with DPE, addition of lithium chloride and THF, and then polymerization of methyl methacrylate at –78 °C to form the two end blocks [119].

However, the use of 1,1-diphenylalkyllithium initiators for initiation of styrene and diene monomers is more problematic, since addition of these monomers forms much less stable styryl and dienyl carbanions (see Table 2). Lee [109] has investigated and optimized these crossover reactions from polymeric 1,1-diphenylalkyllithiums to dienes and styrenes (see Scheme 8).

4.1
Crossover Reactions of Polymeric 1,1-Diphenylalkylithiums to Dienes

In accord with the previously described work of Morton and Fetters [49] and Guzman and coworkers [91] for low molecular weight 1,1-diphenylalkyllithiums, B. Lee [109] found that the crossover reactions of PS-DPELi (the adduct of PSLi with DPE) to both isoprene and butadiene monomers were completed in a few minutes in cyclohexane. Even at 10 °C, the crossover reaction to isoprene was completed in approximately 2 min. In accord with expectations, the SEC traces of the resulting block copolymers were clean and monomodal. It was also found that PS-DPELi was an efficient initiator, i.e., M_n(calc)-out≈M_n(obs)-out, and very narrow molecular weight distributions (M_w/M_n=1.01) were obtained for the block copolymers resulting from these crossover reactions, even for number average molecular weights as low as 2900 for the polybutadiene block. Thus, both simple and polymeric 1,1-diphenylalkyllithium initiators are useful for the synthesis of well-defined polydiene block segments with predictable molecular weights and narrow molecular weight distributions.

4.2
Crossover Reactions of Polymeric 1,1-Diphenylalkylithiums to Styrene

In contrast to the results observed for crossover reactions to diene monomers, the crossover reactions of PS-DPELi to styrene monomer required 1–2 h for completion in benzene and in cyclohexane. However, it was found that PS-DPELi was an efficient initiator, i.e., M_n(calc)-out≈M_n(obs)-out, for crossover to styrene monomer, provided that the number average molecular weight of the formed polystyrene block was=4000.

The apparent rate constants (k_2') (see Eq. 20) for the crossover reaction of PS-DPELi to styrene monomer were determined from the initial slope of the pseudo first order plot of the decay of the absorption peak for PS-DPELi:

$$\frac{-d[S]}{dt} = k_2'[PSDPELi]_{tot}^{1/2}[S] \qquad (20)$$

These apparent crossover rate constants are listed in Table 4 and compared with analogous apparent rate constants for propagation (k_p') of styrene monomer (Eq. 21):

$$\frac{-d[S]}{dt} = k_p'[PSLi]_{tot}^{1/2}[S] \qquad (21)$$

An analogous rate constant determined by Yamagishi and Szwarc [121] is also included for comparison.

The data in Table 4 show that the rate of crossover from PS-DPELi to styrene monomer is slower by at least a factor of ten compared to the propagation rate constant for styrene under the same conditions. This discrepancy provides an explanation for the fact that somewhat broader molecular weight distributions are obtained for polystyrene blocks initiated by PS-DPELi compared to the corresponding polydiene blocks. This also explains why it is necessary to have a minimum block molecular weight for the growing polystyrene block (M_n=4,000) since residual PS-DPELi is observed for formation of lower molecular weight blocks because of the slower rate of crossover relative to propagation.

Table 4. Comparison of the apparent first order rate constants, k_2', for crossover from PSDPELi to styrene monomer with styrene propagation rate constants, k_p', in cyclohexane and in benzene [109]

Solvent	k_2' [$M^{-1/2} s^{-1}$]	k_p' [$M^{-1/2} s^{-1}$]	k_2'/k_p'
Cyclohexane[a,b]	3.7×10^{-4}	3.1×10^{-3} [c]	0.12
Benzene[a,b]	$6.6 \times 10^{-4} (7.7 \times 10^{-4})$ [d]	9.4×10^{-3} [e]	0.07

[a] M_n (PS initial)=1500; [PSDPELi]=1.0×10^{-3} mol/l; M_n (PS 2nd block)=10,000.
[b] At 25 °C.
[c] Obtained from [120].
[d] Obtained from [121].
[e] Obtained from [122].

5
Copolymerization of 1,1-Diphenylethylene with Styrenes and Dienes

Although 1,1-diphenylethylene does not hompolymerize, it will copolymerize with styrene and diene monomers. The anionic copolymerization behavior of 1,1-diphenylethylene with these monomers can be described in terms of the Mayo-Lewis copolymerization equation (Eq. 22) [123, 124]:

$$\frac{dM_1}{dM_2} = \frac{[m_1](r_1[m_1] + [m_2])}{[m_2](r_2[m_2] + [m_1])} \tag{22}$$

where M_i refers to the amount of monomer i incorporated into the copolymer at a time t, [m_i] refers to the concentration of monomer i in the feed, r_1 and r_2 are the monomer reactivity ratios for monomer 1 and monomer 2, respectively, as defined in Eq. (23) using the kinetic parameters described in Eqs. (24–27):

$$r_1 = k_{11}/k_{12} \qquad r_2 = k_{22}/k_{21} \tag{23}$$

$$\text{wwww}M_1^\ominus + M_1 \xrightarrow{k_{11}} \text{wwww}M_1{-}M_1^\ominus \tag{24}$$

$$\text{wwww}M_1^\ominus + M_2 \xrightarrow{k_{12}} \text{wwww}M_1{-}M_2^\ominus \tag{25}$$

$$\text{wwww}M_2^\ominus + M_2 \xrightarrow{k_{22}} \text{wwww}M_2{-}M_2^\ominus \tag{26}$$

$$\text{wwww}M_2^\ominus + M_1 \xrightarrow{k_{21}} \text{wwww}M_2{-}M_1^\ominus \tag{27}$$

It is assumed that the reactivity of the chain end depends only on the last unit added to the chain end, i.e., there are no penultimate effects.

The Mayo-Lewis equation is general for all chain reaction polymerizations regardless of whether the mechanism involves anionic, cationic, radical, or organometallic reactive chain species; furthermore, it is independent of the mode of termination of chain growth.

Because DPE cannot homopolymerize, its monomer reactivity ratio is zero, i.e., $k_{22}=0$. When 1,1-diphenylethylene copolymerizes with styrene or dienes, an alternating-type copolymer is obtained ($r_1r_2=0$) [125].

5.1
Copolymerization with Styrenes

A study of the copolymerization of 1,1-diphenylethylene and styrene (M_1) showed that an almost perfectly alternating copolymer was prepared in toluene at 30 °C with n-butyllithium as initiator ($r_1=0.44$, see Table 5) [126]. When the temperature was changed to 0 °C or 50 °C, the reactivity ratio r_1 changed only slightly to 0.47 and 0.55, respectively. Thus, it seems that the reactivity ratio r_1 is relatively independent of temperature. These results show that, with respect to addition to poly(styryl)lithium, 1,1-diphenylethylene is more reactive than styrene by a factor of approximately two in hydrocarbon solution. This is the same relative reactivity for DPE vs styrene that was deduced from the rate constants for addition of n-butyllithium [23, 41] (see Sect. 2.1.2). When copolymerizations of styrene and DPE were performed in THF, the monomer reactivity ratio for styrene was reduced as shown in Table 5. Thus, in THF 1,1-diphenylethylene is 4.8 times more reactive than styrene with respect to addition to poly(styryl)lithium chain ends. It is noteworthy that the softening points of the resulting alternating styrene-DPE copolymers were in the range of 169–175 °C when the ratio of styrene/DPE was in the range of 1.1–1.3 [126]. Trepka [127] reported that styrene/DPE copolymers with softening temperatures of 182 °C could be obtained. When DPE/styrene end blocks were incorporated into ABA triblock copolymers with butadiene center blocks, thermoplastic elastomers with significantly greater tensile properties and a higher upper use temperature were obtained.

Table 5. Styrene monomer reactivity ratios for copolymerizations of styrene (M_1) with 1,1-diphenylethylene using n-butyllithium as initiator [126]

Solvent	Temperature, °C	r_1
Toluene	0	0.47
	30	0.44
	50	0.55
Benzene	30	0.71
Hexane	30	0.63
THF	30	0.13

Table 6. Monomer reactivity ratios (r_1) for copolymerization of substituted styrenes (M_1) with 1,1-diphenylethylene using n-butyllithium as initiator [125, 129, 130]

M_1	r_1 in THF (0 °C)	r_1 in benzene (40 °C)
Styrene	0.13	0.44
o-Methylstyrene	0.17	2.7
o-Methoxystyrene	~0	20
p-Methoxystyrene	~0	<0.3
m-Methoxystyrene	0.26	8.7
m-Divinylbenzene	1.2[a]	2.5[b]
p-Divinylbenzene	2.8[a]	16[b]

[a] –78 °C.
[b] Toluene at –20 °C.

Ureta et al. [128] reported that the kinetics of copolymerization of styrene and DPE with sodium as counterion in THF were complicated by the reversibility of the addition of styrene to the polymeric 1,1-diphenylmethyl carbanion as shown in Eq. (28):

$$\text{wwww}M_2^\ominus + M_1 \underset{k_r}{\overset{k_{21}}{\rightleftarrows}} \text{wwww}M_2-M_1^\ominus \qquad (28)$$

The apparent rate constant was k_{21}=0.5–0.7 l/mol s and the equilibrium constant was K_{21}=5×10^{-2} l/mol.

Yuki and Okamoto [129] investigated the copolymerization of ring-substituted styrenes with 1,1-diphenylethylene and the results are summarized in Table 6 [125]. Highly alternating copolymers were obtained for all monomers in THF. However, in benzene the *ortho-* and *meta*-substituted styrenes exhibited much larger monomer reactivity ratios. These results suggest that, with these substitution patterns, steric effects of the substituents with the incoming 1,1-diphenylethylene retard crossover relative to addition of the substituted styrene monomer to the chain end. The relatively large monomer reactivity ratio for o-methoxystyrene in benzene has been ascribed to coordination of the methoxy group of the incoming monomer with the lithium counterion at the chain end as illustrated schematically in **14** below. Consistent with intramolecular coordination is the fact that highly *isotactic* poly(o-methoxystyrene) is obtained in toluene at low temperatures [131].

Hatada and coworkers [130, 132] have investigated the copolymerization of divinylbenzenes with 1,1-diphenylethylene and the monomer reactivity ratios are listed in Table 6. Using ratios of DVB/DPE=0.5 in THF at –78 °C, soluble copolymers could be obtained with compositions of DVB/DPE=1.2–1.4. The monomer reactivity ratios in Table 6 suggest that a more highly alternating polymer structure will be obtained with *meta*-DVB compared to *para*-DVB. The resulting polymers were reacted with butyllithium at –78 °C in THF to form anionic sites which were used to initiate graft copolymerization of methyl methacrylate.

5.2
Copolymerization with Dienes

Studies of the copolymerizations of 1,1-diphenylethylene and dienes showed rather different behavior compared with the copolymerizations of styrene and 1,1-diphenylethylene [125, 133–136]. The monomer reactivity ratios for copolymerizations of dienes with DPE are shown in Table 7. When butadiene was copolymerized with 1,1-diphenylethylene in benzene at 40 °C with n-butyllithium as initiator, the monomer reactivity ratio for butadiene, r_1, was 54; this means that the addition of butadiene to the butadienyl anion is 54 times faster than addition of 1,1-diphenylethylene to the butadienyl anion [133]. This unreactivity of poly(butadienyl)lithium towards addition to DPE was also observed in studies of end-capping of poly(butadienyl)lithium with DPE in hydrocarbon solution (see Sect. 3.3) [109, 111]. Because of this unfavorable monomer reactivity ratio, few DPE units would be incorporated into the co-

Table 7. Monomer reactivity ratios (r_1) for the copolymerization of dienes (M_1) and 1,1-diphenylethylene using n-butyllithium as initiator [125, 133–136]

Monomers	r_1	
	THF[a]	Benzene[b]
Butadiene	0.13	54
Isoprene	0.11	37
2,3-Dimethylbutadiene	~0[c]	0.23

[a] At 0 °C.
[b] At 40 °C.
[c] At 22 °C.

polymer in hydrocarbon solution with lithium as counterion. It is interesting to note that this unreactivity of butadienyl carbanions toward DPE is unique to lithium; the monomer reactivity ratios for butadiene in benzene were 0.71 and 0.10 with sodium and potassium as the counterions [125, 133]. Thus, with sodium or potassium as counterions, DPE is more reactive than butadiene with respect to addition to butadienyl carbanionic chain ends. When THF was used as solvent at 0 °C with lithium as counterion, r_1 decreased to 0.13. Thus, in THF a highly alternating copolymer structure would be expected.

This lack of copolymerization reactivity of DPE was also observed for the copolymerization of 1,1-diphenylethylene and isoprene [125, 134]. As shown in Table 7, when isoprene was copolymerized with DPE in benzene using *n*-butyllithium as initiator, the monomer reactivity ratio was 37, which indicates that the addition of isoprene to the isoprenyllithium chain end is 37 times faster than the addition of 1,1-diphenylethylene. The unreactivity of isoprenyl carbanions toward DPE is unique to lithium; the monomer reactivity ratios for isoprene in benzene were 0.38 and 0.05 with sodium and potassium as the counterions [125, 134]. When THF was used as the solvent at 0 °C, r_1 decreased to 0.11 with lithium as counterion.

Surprising results were obtained for the copolymerization of 2,3-dimethylbutadiene and 1,1-diphenylethylene [125, 135, 136]. When these two monomers were copolymerized in benzene at 40 °C with *n*-butyllithium as initiator, the reactivity ratio for 2,3-dimethylbutadiene, r_1, was 0.23, while in THF at 22 °C, r_1 decreased to about 0 (see Table 7). These results show that, with increased steric requirements in the 1,3-diene, the r_1 value decreases dramatically and DPE is preferentially added compared to the hindered diene, 2,3-dimethylbutadiene. One of the main factors affecting the reactivity of organolithium compounds is the degree of aggregation; it has been found that less aggregated organolithium compounds are more reactive [3, 19]. In general, more sterically hindered organolithiums are less aggregated. Therefore, it is possible that poly(2,3-dimethylbutadienyl)lithium has either a lower degree of aggregation or weaker aggregation than poly(butadienyl)lithium or poly(isoprenyl)lithium; this would explain the dramatic increase in reactivity in hydrocarbon solution.

One of the most surprising aspects of the copolymerizations of dienes with 1,1-diphenylethylene is the diene stereochemistry of the highly alternating copolymers formed in tetrahydrofuran with butyllithium as initiator [125, 133]. In general, high 1,4-diene microstructure is only observed for lithium as counterion in hydrocarbon solution [3, 81]. However, diene/DPE copolymers exhibit much higher 1,4-diene microstructures in tetrahydrofuran than the corresponding diene homopolymers under the same conditions as shown in Table 8 [125, 133]. For example, while polybutadiene prepared in THF contains 81% 1,2-enchainment, the corresponding DPE copolymer exhibits 70% 1,4-enchainment. Similarly, while polyisoprene prepared in THF contains 71% 3,4-enchainment, the corresponding DPE copolymer exhibits 83% 1,4-enchainment. A similar trend was observed for 2,3-dimethyl-1,3-butadiene where the

Table 8. Diene microstructure in homopolymers and alternating copolymers with 1,1-diphenylethylene obtained using n-butyllithium as initiator in THF [125, 133]

Diene	Diene homopolymer microstructure (mol %)			Alternating copolymer microstructure (mol %)		
	1,2-	1,4-	3,4-	1,2-	1,4-	3,4-
1,3-Butadiene	82	18	–	30	70	–
Isoprene	29	~0	71	~0	83	17
2,3-Dimethyl-1,3-butadiene	49	51	–	0	100	–

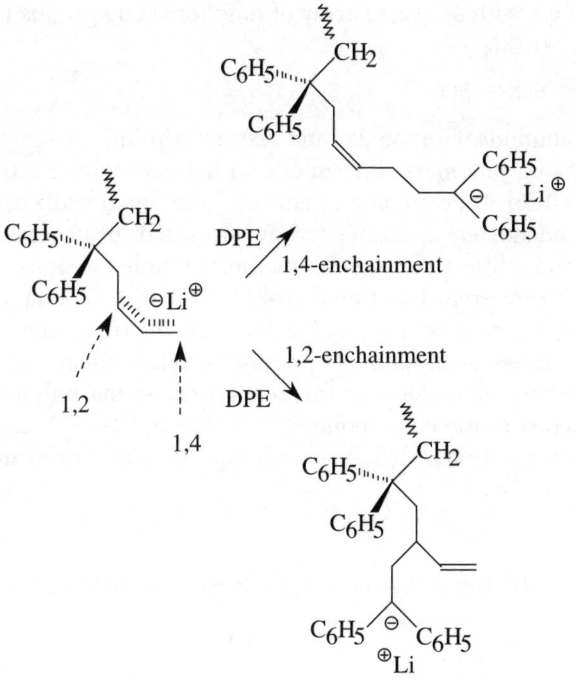

Scheme 9.

copolymer contains essentially all 1,4-units. Yuki [125] has proposed that the microstructure of diene units in the corresponding alternating DPE copolymer is controlled by steric factors. The competing reactions are illustrated in Scheme 9. It would be reasonable to expect that the diphenylmethyl group in the penultimate unit would tend to hinder sterically approach of the incoming monomer to form the 1,2-diene unit relative to attack at the less hindered terminal methylene unit. In addition, this steric congestion is exacerbated by the steric requirements of the incoming 1,1-diphenylethylene unit in this alternating copolymerization.

6
Anionic Functionalization Chemistry of 1,1-Diphenylethylenes

6.1
Introduction to Anionic Functionalization Reactions

The methodology of living anionic polymerization, especially alkyllithium-initiated polymerizations of styrene and diene monomers, is particularly suitable for the synthesis of functionalized polymers with well-defined structures [3, 137]. Since these living polymerizations generate stable, anionic polymer chain ends (PLi) when all of the monomer has been consumed, post-polymerization reactions with a variety of electrophilic species (X-Y) can be used to generate polymers with a diverse array of functional end groups (P-X) as illustrated in Eq. (29) [8]:

$$PLi + X\text{-}Y \rightarrow P\text{-}X + LiY \tag{29}$$

The literature abounds with tabular and text descriptions of functionalization reactions for living polymeric carbanions with a variety of electrophilic species [2, 19, 138–140]. Unfortunately, many of these functionalization reactions have not been adequately characterized or optimized for general utility [3, 8, 137]. Furthermore, different, specific functionalization reactions are generally required for each different functional group. Thus, it is necessary to develop, analyze, and optimize new procedures for each different functional group. One factor which has been found to promote efficient chain-end functionalization for a variety of systems is the conversion of the polymeric organolithium chain ends to the corresponding 1,1-diphenylalkyllithium chain ends by reaction with 1,1-diphenylethylene (see Eq. 13) as discussed in the following section.

6.2
Applications of 1,1-Diphenylethylene in Specific Functionalization Reactions

6.2.1
Carbonation

The limitations of the anionic polymerization methodology of using specific functionalization reactions to prepare chain-end functionalized polymers can be clearly illustrated by the carbonation reaction. The carbonation of polymeric carbanions using carbon dioxide is one of the most useful and widely used functionalization reactions. However, the direct carbonation of polymeric organolithium compounds is not an efficient reaction [3, 137]. For example, when carbonations with high-purity, gaseous carbon dioxide are carried out in benzene solution at room temperature using standard high vacuum techniques, the carboxylated polymers are obtained in only 27–66% yields for poly(styryl)lithium, poly(isoprenyl)lithium, and poly(styrene-b-isoprenyl)lithium [141]. The functionalized polymer is contaminated with dimeric ketone (23–27%) and trimeric alcohol (7–50%) as shown in Eq. (30):

$$\text{PLi} \xrightarrow{CO_2} \xrightarrow{H_3O^+} \text{PCO}_2\text{H} + \text{P}_2\text{CO} + \text{P}_3\text{COH} \tag{30}$$

In order to optimize specific anionic functionalization reactions such as carbonation with carbon dioxide, the effect of chain end structure (stability and steric requirements) has often been investigated. The steric and electronic nature of the anionic chain end and the chain-end aggregation can be modified by reaction with 1,1-diphenylethylene as shown in Eq. (13). When the direct carbonation is effected in benzene at room temperature with the diphenylalkyllithium species formed by addition of poly(styryl)lithium to 1,1-diphenylethylene (Eq. 31), the carboxylated polymer can be isolated in 98% yield compared to only a 47% yield for poly(styryl)lithium without end-capping under the same conditions [141]:

$$\text{PSLi} + \text{CH}_2=\text{C}(\text{C}_6\text{H}_5)_2 \longrightarrow \text{PSCH}_2\text{C}(\text{C}_6\text{H}_5)_2\text{Li} \xrightarrow[2)\ H_3O^+]{1)\ CO_2} \text{PSCH}_2\text{C}(\text{C}_6\text{H}_5)_2\text{CO}_2\text{H} \tag{31}$$

6.2.2
Sulfonation

Another dramatic example of the usefulness of DPE end-capping of polymeric organolithium compounds to promote efficient functionalization reactions is the sulfonation reaction using sultones [142]. A careful examination of the functionalization of poly(styryl)lithium with 1,3-propane sultone showed that the corresponding sulfonated polymer (see Eq. 32) was obtained in maximum yields of only 30% and 53% in benzene or tetrahydrofuran, respectively [142]:

$$\text{PSLi} + \overset{O}{\underset{SO_2}{\bigcirc}} \longrightarrow \text{PSCH}_2\text{CH}_2\text{CH}_2\text{SO}_3^-\text{Li}^+ \tag{32}$$

It was proposed that the low yields of sulfonated polymer resulted from the competing metalation reaction of poly(styryl)lithium with the acidic α-hydrogens of the sultone as shown in Eq. (33) [143]:

$$\text{PSLi} + \overset{O}{\underset{SO_2}{\bigcirc}} \longrightarrow \text{PSH} + \overset{O}{\underset{SO_2}{\bigcirc}}{}^{\ominus}\text{Li}^{\oplus} \tag{33}$$

When poly(styryl)lithium was end-capped with 1,1-diphenylethylene prior to sulfonation with the sultone, the corresponding sulfonated polymer was obtained in 93% yield (Scheme 10). Both the increased steric requirements of the chain end and the more stable 1,1-diphenylalkyllithium chain end appar-

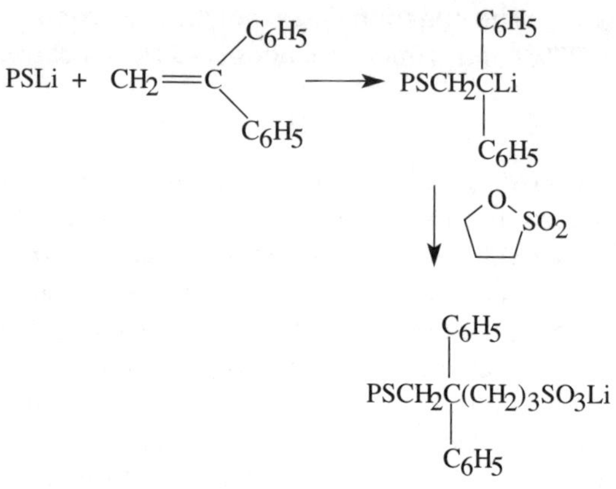

$$\text{PSLi} + \text{CH}_2=\text{C}\begin{smallmatrix}\text{C}_6\text{H}_5\\ \\ \text{C}_6\text{H}_5\end{smallmatrix} \longrightarrow \text{PSCH}_2\underset{\underset{\text{C}_6\text{H}_5}{|}}{\overset{\overset{\text{C}_6\text{H}_5}{|}}{\text{C}}}\text{Li}$$

$$\downarrow \text{\large{\textit{(cyclic sultone)}}} \;\; \begin{smallmatrix}O\\ \diagup \diagdown \\ \;\;\;\;\; SO_2\end{smallmatrix}$$

$$\text{PSCH}_2\underset{\underset{\text{C}_6\text{H}_5}{|}}{\overset{\overset{\text{C}_6\text{H}_5}{|}}{\text{C}}}(\text{CH}_2)_3\text{SO}_3\text{Li}$$

Scheme 10.

ently contributed to the decreased reactivity toward the competing metalation side reaction compared with ring-opening to form the functionalized polymer. These procedures have been applied to the synthesis of α,ω-disulfonatopolystyrenes [144]; the sulfonation yields were reported to be >95%.

6.2.3
Epoxide-Functionalized Macromonomers

The reaction of poly(styryl)lithium (PSLi) with epichlorohydrin (EPC) in benzene produced the corresponding ω-epoxide-functionalized polystyrene, **15**, in only 9% yield (Eq. 34); the main product corresponded to dimeric species (70%) and ring-opened products [145]:

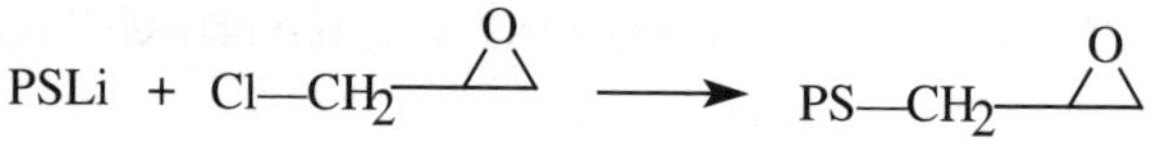

$$\text{PSLi} + \text{Cl}-\text{CH}_2-\overset{O}{\overset{\triangle}{}} \longrightarrow \text{PS}-\text{CH}_2-\overset{O}{\overset{\triangle}{}} \quad\quad\quad (34)$$
$$\mathbf{15}$$

The most efficient synthesis of an epoxide-functionalized macromonomer (97% yield, no dimer) was obtained by first end-capping poly(styryl)lithium with 1,1-diphenylethylene as shown in Scheme 11. The decreased reactivity and increased steric requirements of the diphenylalkyllithium chain end promoted selectivity and decreased the formation of dimer in this epoxide functionalization reaction.

Because of the limitations of specific electrophilic functionalization reactions, there has been a need to develop general functionalization reactions which proceed efficiently to introduce a variety of different functional groups. One of the most useful general functionalization reactions which has been developed is the use of ring-substituted 1,1-diphenylethylenes as described in the following section.

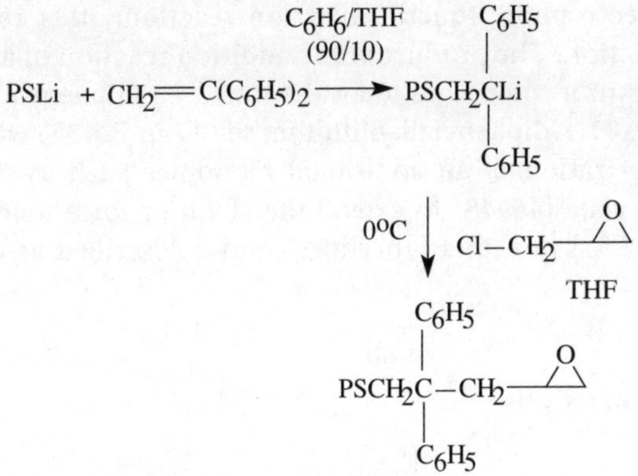

Scheme 11.

6.3
General Aspects of Anionic Functionalization Chemistry with 1,1-Diphenylethylenes

The reaction of polymeric organolithium compounds with substituted 1,1-diphenylethylene derivatives (16) has been shown to be an excellent methodology for anionic synthesis of chain-end functionalized polymers (18) (Eq. 35)

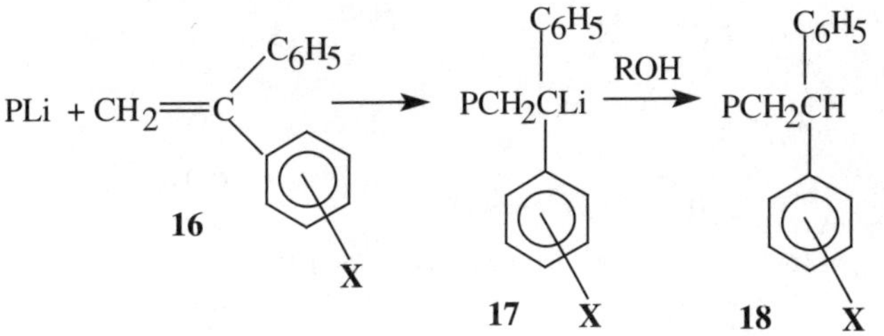

$$\text{PLi} + \text{CH}_2=\text{C} \begin{array}{c} \text{C}_6\text{H}_5 \\ \text{(C}_6\text{H}_4\text{X)} \end{array} \longrightarrow \text{PCH}_2\text{CLi}\begin{array}{c}\text{C}_6\text{H}_5\\\text{(C}_6\text{H}_4\text{X)}\end{array} \xrightarrow{\text{ROH}} \text{PCH}_2\text{CH}\begin{array}{c}\text{C}_6\text{H}_5\\\text{(C}_6\text{H}_4\text{X)}\end{array} \tag{35}$$

because:
1. These addition reactions are simple and quantitative
2. Only monoaddition, i.e., no oligomerization has been reported using stoichiometric concentrations
3. The rate and efficiency of the crossover reaction can be monitored by ultraviolet-visible spectroscopy
4. Copolymerization of substituted 1,1-diphenylethylene derivatives with other monomers will result in polymers with multiple functional groups along the polymer chain
5. These addition reactions take place readily in hydrocarbon solution at room temperature and above
6. A variety of substituted 1,1-diphenylethylenes with functional groups on the aromatic ring can be prepared readily [3, 137, 146, 147].

Unlike most electrophilic functionalization reactions, this reaction is not a termination reaction. The product of the addition reaction of a simple or polymeric organolithium compound to a substituted 1,1-diphenylethylene is a carbanionic species (1,1-diphenylalkyllithium; see **17** in Eq. 35) which can initiate anionic polymerization of an additional monomer such as isoprene [49] or methyl methacrylate [46–48] to extend the chain or form a new block as discussed in Sect. 2.3. Thus, this procedure can be described as a *living functio-*

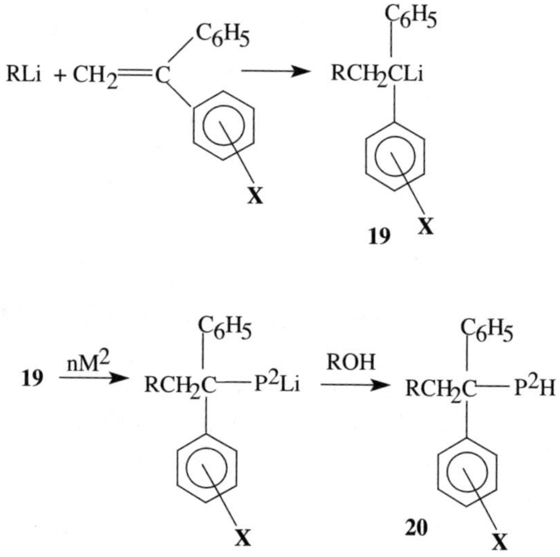

Scheme 12.

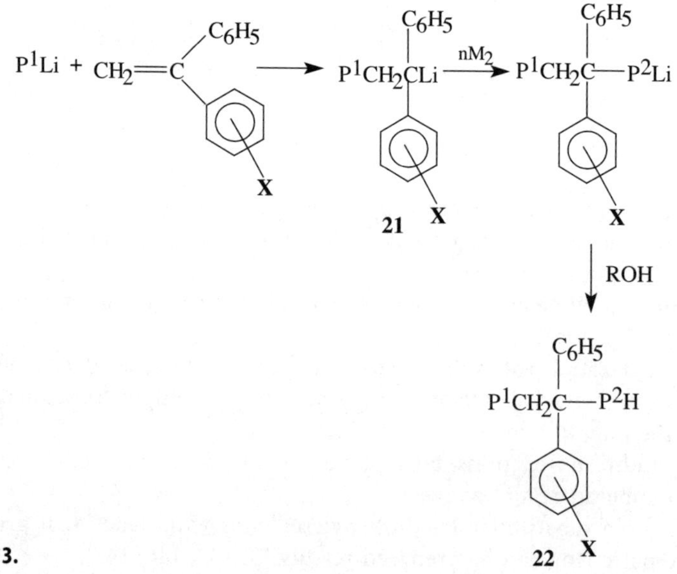

Scheme 13.

nalization reaction. Consequently, this method can be used to prepare polymers with functional groups at the initiating end (α) of the polymer chain (**20** in Scheme 12) by reaction of a substituted 1,1-diphenylethylene with an alkyllithium compound (RLi) to form the corresponding functionalized initiator, **19**. Functional groups can also be inserted within the polymer chain (**22** in Scheme 13), for example, at the junction between two blocks. Thus, the 1,1-diphenylethylene functionalization methodology provides a versatile, general anionic functionalization procedure with which one can rationally design and place functional groups at essentially any position in a polymer molecule. Furthermore, this methodology can be used to synthesize polymers with two or more functional groups by utilizing 1,1-diphenylethylene derivatives which contains two or more functional groups. By analogy with the monofunctional chemistry described in Eq. (35) and Schemes 12 and 13, these functional groups can be located at the initiating chain end, within the polymer chain or at the terminating chain end as exemplified in Eq. (36) for terminal difunctionalization, where X and Y can either be the same or different functional groups [148]:

$$\text{PLi} + \text{CH}_2=\text{C} \longrightarrow \text{PCH}_2\text{CLi} \xrightarrow{\text{ROH}} \text{PCH}_2\text{CH} \quad \quad (36)$$

23

The α,α- or ω,ω-difunctional products of these functionalization reactions can behave as macromolecular monomers for condensation-type copolymerization reactions to form comb-type graft copolymers [149–151].

The following sections describe the applications of substituted 1,1-diphenylethylene chemistry for the preparation of phenol, amino, and carboxyl functionalized polymers as well as for the preparation of polymers with aromatic functional groups which provide fluorescent labels. These examples show that anionic functionalization chemistry utilizing substituted 1,1-diphenylethylenes as outlined herein provides a versatile and general functionalization methodology for quantitative introduction of functional groups at the chain ends and within the polymer chain by design.

6.3.1
Protecting Groups

Because many functional groups of interest (e.g., hydroxyl, carboxyl, amino) are not stable in the presence of either simple or polymeric organolithium reagents, it is generally necessary to use suitable protecting groups for the substituted 1,1-diphenylethylene functionalizing agents [152]. Fortunately, protecting groups for a wide variety of functional groups have been developed for applications in anionic polymerization chemistry [79, 153–157]. However, many of the previous applications of these protecting groups involved low temperatures (e.g., –78 °C) in polar solvents such as tetrahydrofuran (THF) and counterions other than lithium [155, 156]. It was thus necessary to determine the suitability of many of these protecting groups for use with organolithium compounds in hydrocarbon solution at room temperature and above, i.e., under conditions which promote chain end stability and high 1,4-polydiene stereospecificity.

6.3.2
Phenol Functionality

The first application of this methodology was directed to the preparation of phenol-terminated polymers [148, 158]. The *tert*-butyldimethylsilyl-protecting group was chosen for the aromatic hydroxyl group based on the previous results reported by Nakahama and coworkers [155, 156, 159]. The addition of poly(styryl)lithium to 1-(4-*tert*-butyldimethylsiloxyphenyl)-1-phenylethylene (24) (0.2 molar excess) in benzene at room temperature was monitored by the appearance of a ultraviolet-visible absorption at 406 nm; complete addition required three days at room temperature.

The corresponding reaction with DPE was completed in a matter of hours [109]. The reduced rate of addition to the siloxyl-substituted DPE is consistent with the Hammett ρ value of +1.8 for these reactions [95] as discussed in Sect. 3.2.1. The phenol-end-functionalized polystyrenes were obtained in >99% yield after hydrolysis with 1% HCl in tetrahydrofuran under reflux for 3 h. The efficiency of these functionalization reactions was evaluated by end-group titration, elemental analyses, ^{1}H NMR and ^{13}C NMR analyses, as well as by thin-layer chromatography. All of the available evidence suggested that this is an essentially quantitative functionalization reaction; <1% unfunctionalized polystyrene was formed. It is noteworthy that Heitz and Hocker [148] have carried out similar functionalizations using 1,1-(4,4'-dimethoxyphenyl)ethylene.

6.3.3
Tertiary Amine Functionality

In an analogous fashion, 1-(4-dimethyl-aminophenyl)-1-phenylethylene (**25**) was used to prepare amine- terminated polymers [50, 160, 161].

The reaction of poly(styryl)lithium in benzene solution with **25** (0.2 molar excess) was monitored by ultraviolet-visible spectroscopy at 406 nm; a period of three days was required for complete addition at room temperature. After methanol termination, the ω-dimethylamino-terminated polystyrene was obtained in >99% yields (<1% unfunctionalized polystyrene was detected) [50]. Similarly, the functionalization of poly(butadienyl)lithium in benzene was effected in the presence of THF ([THF]/[PBDLi]=21) to form the corresponding ω-dimethylamino-functionalized polybutadiene with a functionality of 0.94 and high 1,4-microstructure (90.5%) [160]. It is noteworthy that in the absence of THF, no addition of poly(butadienyl)lithium to **25** was detected by UV spectroscopy; this is in accord with the unreactivity of poly(dienyl)lithium compounds with respect to addition to 1,1-diphenylethylenes (see also Sect. 3.3). Telechelic α,ω-bis(dimethylamino)-terminated polystyrene was prepared by first reacting *sec*-butyllithium in benzene with a stoichiometric amount of **25** to form an amine-functionalized initiator, **26**; α-dimethylamino-poly(styryl)lithium, **27**, was prepared by initiating styrene polymerization with **26** as shown in Scheme 14. An aliquot of **27** was removed from the reactor and terminated with methanol to form the corresponding α-dimethylamino-functionalized polystyrene, **28**. This polymer exhibited an amine functionality of 1.2 as determined by amine end group titration. The remainder of the living polymer, **27**, was then functionalized by addition of **25** to form the corresponding α,ω-dimethyl-aminopolystyrene, **29**, after methanol termination as shown in Scheme 14. The SEC chromatograms of both the α-dimethylamino-polystyrene and the α,ω-bis(dimethylamino)polystyrene exhibited narrow molecular weight distributions (M_w/M_n=1.04) [50]. These narrow molecular weight distributions indicate that the amine-functionalized initiator, **26**, initiates styrene polymerization at a rate which is competitive with or faster than propagation [82], a requisite condition for a useful organolithium initiator [80]. Furthermore, there was good agreement between the calculated molecular weight (M_n=19,400) and the observed molecular weight (M_n=20,500) which indicates that the initiator, **26**, is also an efficient initiator for styrene

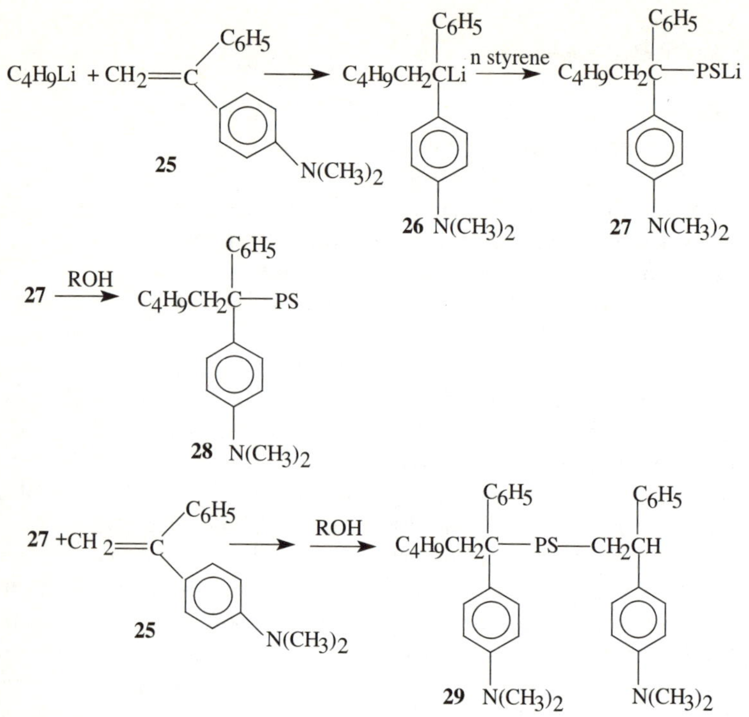

Scheme 14.

polymerization. The amine group functionality of this telechelic polymer was 2.1 as determined by end-group titration [50].

The hydrocarbon-soluble, functionalized initiator, **26**, has also been used to prepare α-4-dimethylaminophenyl-functionalized polybutadiene in cyclohexane with 90.5% 1,4-microstructure [160]. The initiator was efficient for butadiene polymerization since there was good agreement between the calculated molecular weight (M_n=259,500) and the observed molecular weight (M_n=265,000). The molecular weight distribution determined by SEC was narrow (M_w/M_n=1.04), indicating that the initiator initiates at a rate which is competitive with or faster than propagation for butadiene. A functionality of 0.93 for this high molecular weight polybutadiene was determined by UV-visible spectroscopy using the absorptions at λ_{max}=262 and 303 nm. The ability to determine the functionality for high molecular weight polydienes by UV-visible spectroscopy is an added advantage of this 1,1-diphenylethylene functionalization methodology.

Kim and coworkers [162] have recently prepared 1,1-bis[4-dimethylaminophenyl]ethylene and used this difunctional diphenylethylene to prepare α,α'-diaminopolystyrene, ω,ω'-diaminopolystyrene, α,α',ω,ω'-tetraaminopolystyrene, α,α'-diaminopolyisoprene, and α,α',ω,ω'-tetraaminopolyisoprene using analogous chemistry to that shown in Scheme 14.

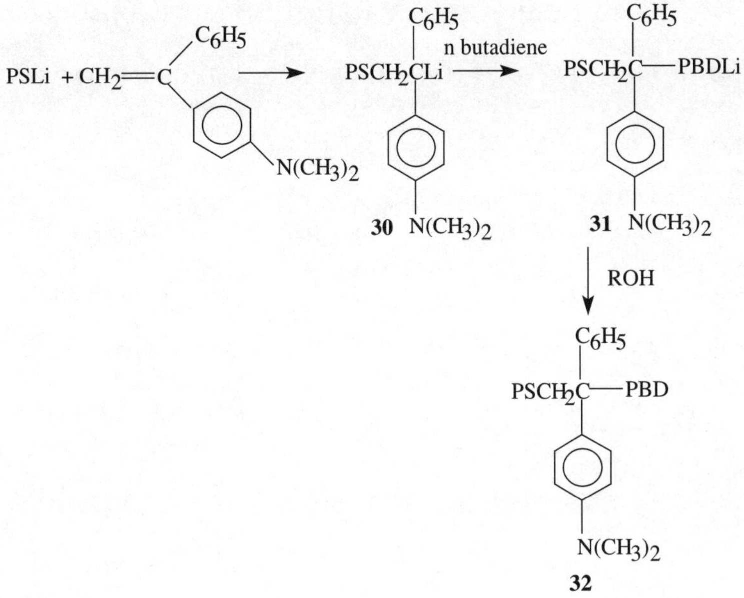

Scheme 15.

The ability to place functional groups within polymer chains was demonstrated by first functionalizing poly(styryl)lithium with **25** to form the functionalized, living polymer, **30**, followed by chain extension with butadiene to form a diblock polymer, poly(styrene-*block*-butadiene), **32**, with the dimethylamino functional group at the interface between the blocks after termination as shown in Scheme 15 [50]. The block copolymer, **32**, exhibited good agreement between the calculated molecular weight (M_n=20,800), the molecular weight from membrane osmometry (M_n=21,300), and end-group titration (M_n=21,900). SEC analysis showed that the molecular weight distribution was narrow (M_w/M_n=1.01). These results show that functionalized diphenylakyllithiums with electron-donating substituents are effective initiators for anionic polymerization of styrene and diene monomers, i.e., polymers with predictable molecular weights and narrow molecular weight distributions are obtained.

The functionalized initiator, **26**, has also been used to initiate polymerization of methyl methacrylate in THF at –78 °C to form the corresponding α-dimethylamino-functionalized poly(methyl methacrylate), **34**, as shown in Scheme 16 [50]. The resulting α-dimethylamino-poly(methyl methacrylate) with 82% syndiotactic diads exhibited a narrow molecular weight distribution as determined by SEC analysis (M_w/M_n=1.06) and the amine functionality was 1.0 as determined by end-group titration. A telechelic α,ω-bis(dimethylamino)poly(methyl methacrylate) (**35**) was prepared by reacting living amine-functionalized poly(methyl methacrylate)lithium enolate (**33**) with a one-half

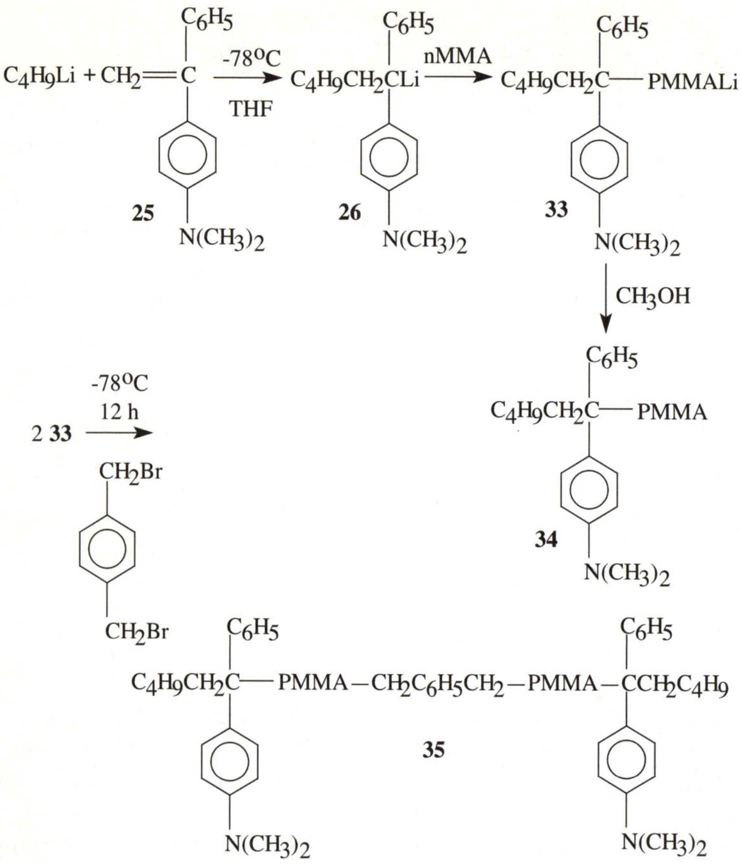

Scheme 16.

molar equivalent of α,α'-dibromoxylene at −78 °C as shown in Scheme 16 [163]. The coupling efficiency to form the telechelic polymer (35) was 94% as determined by SEC analysis. The pure telechelic polymer was isolated by fractionation and exhibited a narrow molecular weight distribution (M_w/M_n=1.06) and the amine functionality was 2.0 as determined by amine end-group titration.

6.3.4
Primary Amine Functionality

α-Primary amine-functionalized polystyrenes, **38**, were synthesized by reacting poly(styryl)lithium with 1-[4-[N,N-bis(trimethylsilyl)amino]phenyl]-1-phenylethylene, **36**, in benzene at room temperature followed by acid-catalyzed hydrolysis and neutralization as shown in Scheme 17 [164]. The isolated amine-functionalized polymer had an amine functionality of 98.5%. Charac-

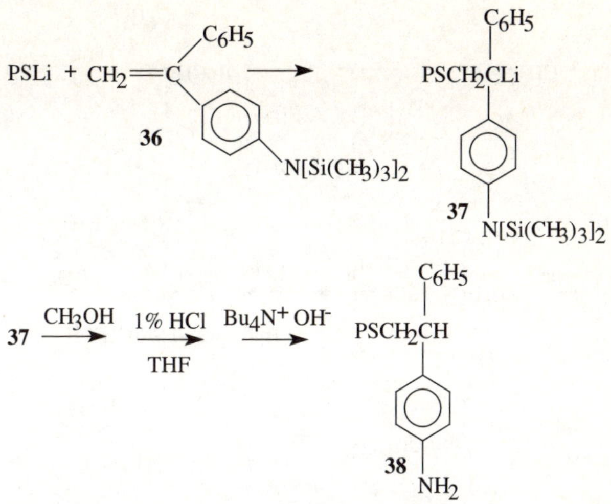

Scheme 17.

terization by ^{1}H NMR, ^{13}C NMR, and FTIR spectroscopy was consistent with the presence of the primary aromatic amine functionality; less than 0.5 wt % of the unfunctionalized polymer was isolated by column chromatography. It is noteworthy that the addition of poly(styryl)lithium to **36** (0.3 molar excess) was complete within 6 h at 25 °C. This can be compared with the rate of addition to 1,1-diphenylethylene, to 1-(4-*tert*-butyldimethylsiloxy-phenyl)-1-phenylethylene (**24**), and 1-(4-dimethylamino-phenyl)-1-phenylethylene (**25**) which required 6 h [109], 3 days [158] and more than 48 h [50], respectively, to go to completion. The comparable reactivity of **36** to 1,1-diphenylethylene is not in accord with expectations based on the Hammett ρ value of +1.8 for addition of poly(styryl)lithium to substituted DPEs [95]. These results suggest that the bis(trimethylsilyl)amine group has reduced electron-donating character compared to the corresponding amine group [164].

Telechelic α,ω-bis(amino)-terminated polystyrene (**41**, Scheme 18) was prepared by first reacting *sec*-butyllithium in benzene with **36** to form the corresponding amine-functionalized initiator, **39** [165]. This addition reaction was complete in 1.5 h, once again consistent with the reduced electron-donating character of the bis(trimethylsilyl)amine group. α-Bis(trimethylsilyl)amino-poly(styryl)lithium (**40**) was prepared by initiating styrene polymerization with initiator **39** in benzene. An aliquot of **40** was removed from the reactor and characterized. The resulting α-bis(trimethylsilyl)aminopolystyrene exhibited a narrow molecular weight distribution (M_w/M_n=1.08) and good agreement between calculated and observed number average molecular weights. These results indicate that the silylamine-functionalized initiator, **39**, initiates styrene polymerization at a rate which is competitive with or faster than propagation and that this initiator is also an efficient initiator for styrene polymerization. The functionalities of the amine-terminated polymers ranged be-

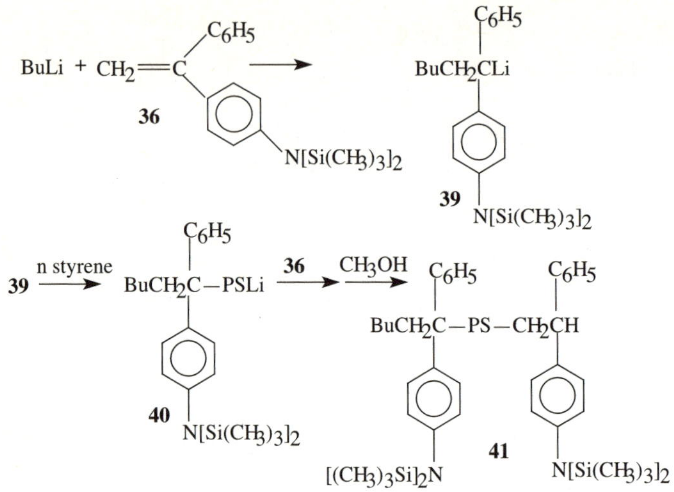

Scheme 18.

tween 89–94% as determined by end-group titration. However, analysis of the functionalized polymers by both TLC and column chromatography indicated that the amount of non-functional polymer was <2 wt % for the monofunctional derivative, 38, and the amount of monofunctional impurity in the difunctional polymer, 41, was approximately 3 wt % [165].

6.3.5
Carboxyl Functionality

The use of substituted 1,1-diphenylethylenes to prepare end-functionalized polymers has also been utilized to prepare carboxyl-functionalized polymers. The carboxyl functionality has been protected using the oxazoline group [166]. However, the oxazoline-substituted 1,1-diphenylethylene (42) was not stable to the anionic chain end at room temperature [167]. Therefore, the functionalization reaction was effected in toluene/THF mixtures (4/1, v/v) at –78 °C to produce the carboxyl-functionalized polymer (44) in quantitative yield after acid hydrolysis as shown in Scheme 19. Quantitative formation of the oxazolyl-functionalized polystyrene (43) was determined by elemental analysis of the polymer.

The carboxyl group has also been protected using the diisopropylamide derivative [168]. It was found that the diisopropylamide group in 45 was not stable to the organolithium chain ends at room temperature [169]. Therefore, the functionalization reaction was effected in toluene/THF mixtures, (4/1, v/v) at –78 °C to produce the amide-functionalized polystyrenes in 92–100% yields as shown in Scheme 20. Although it was somewhat difficult to hydrolyze the amide, heating under reflux in toluene with toluenesulfonic acid was found to be effective in generating the carboxyl-functionalized polymers [169].

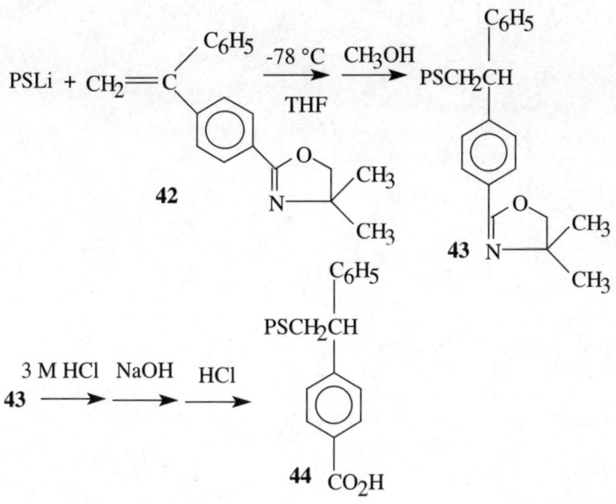

Scheme 19.

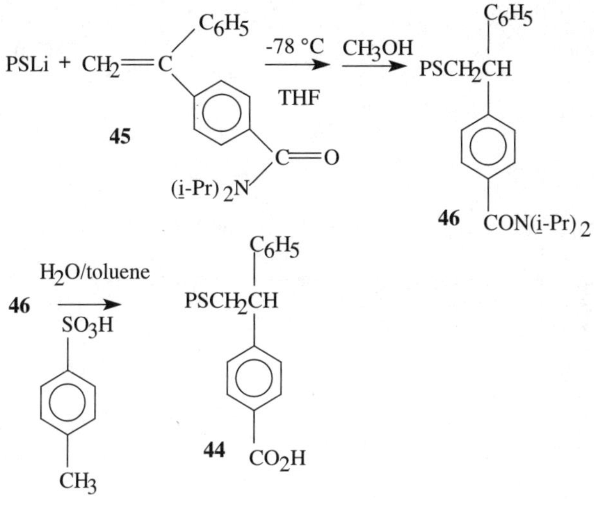

Scheme 20.

6.3.6
Condensation Macromonomers: Diphenol Functionality

Another advantage of the 1,1-diphenylethylene functionalization methodology is that it can be used to prepare condensation macromonomers, **47**, which are polymers with two polymerizable functional groups at one chain end.

This type of macromonomer can participate in step-growth (condensation) polymerization with other difunctional monomers to form model comb-type,

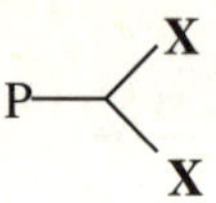

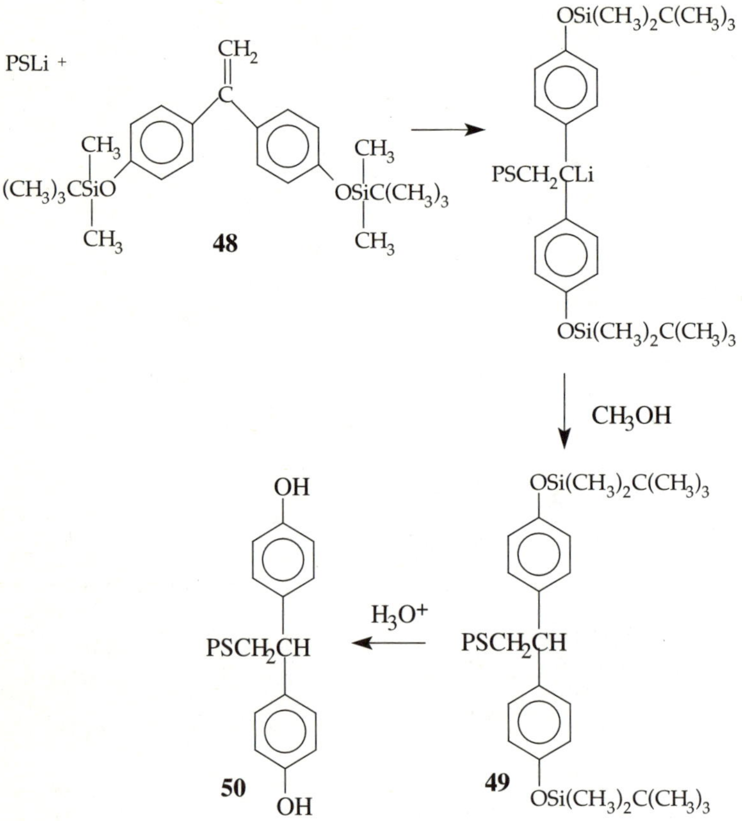

Scheme 21.

branched condensation copolymers [149–151]. ω,ω′-Diphenolpolystyrenes, **50**, were synthesized in quantitative yields by reacting poly(styryl)lithium with 1,1-bis(4-*tert*-butyldimethylsiloxyphenyl)ethylene, **48**, followed by methanol termination and hydrolysis with dilute acid as shown in Scheme 21 [170]. No unfunctional polystyrene was detected by TLC analysis of the functionalized polymers (**49**, **50**) which were also characterized by end-group titration, UV-visible, ^{1}H and ^{13}C NMR spectroscopy.

In accord with the Hammett ρ value of +1.8 for additions of poly(styryl)-lithium with substituted 1,1-diphenylethylenes [95], the rate of addition of

PSLi to **48** was retarded relative to DPE. The addition reaction in benzene required 3–5 days for completion as monitored by UV-visible spectroscopy. The addition was accelerated in the presence of small amounts of THF ([THF]/[Li]=30) such that addition was complete in 12 h. Using the rate constant for addition of PSLi to **48** [$1.7 \times 10^{-3}\,M^{-1/2}\,s^{-1}$], the sigma value ($\sigma$) for the *tert*-butyldimethylsiloxy substituent was calculated to be –0.3 compared to the σ value of –0.46 calculated for this substituent determined from the kinetics of addition of PSLi to **24** under the same conditions [170]. The reduced effectiveness with two substituents is consistent with the X-ray structural evidence for diethyl ether solvated diphenylmethyllithiums which indicates that one of the phenyl rings is distorted by an angle of 37° from coplanarity with the other ring and the diphenylmethyl carbon [52] as discussed in Sect. 2.2.

6.3.7
Trimethylsilyl Group Labeling for Molecular Weight Measurements

Nakahama and coworkers [171–173] have developed the bis(4,4'-trimethylsilyl)-substituted 1,1-diphenylethylene, **51**, to generate the corresponding functional initiators which were used to determine the molecular weights of the resultant polymers by ^{1}H NMR spectroscopy. For example, the polymerization of 4-cyanostyrene was initiated with the adduct of **51** with butyllithium (**52**), as well as the dipotassium dimer of **51** (**54**, Eq. 37),

$$2\,\text{NaphK} + 2\,\mathbf{51} \longrightarrow \mathbf{54}\;\;\text{KCCH}_2\text{—CH}_2\text{CK} \qquad (37)$$

(with structure **54** bearing two 4-(CH$_3$)$_3$Si-substituted phenyl groups on each central carbon)

to form the corresponding labeled polymer (**53**) as shown in Scheme 22. All of these reactions were performed at –78 °C in THF. In general, the corresponding polymers exhibited good agreement between calculated and observed number average molecular weights (determined by ^{1}H NMR) and narrow molecular weight distributions. These results indicate that the trimethylsilyl-substituted 1,1-diphenylalkyl anion initiators are efficient and initiate polymerization at rates which are competitive with propagation. The authors have utilized the sharp signal from the 18 (or 36) trimethylsilyl protons to determine molecular weights as high as 122,000 [172]; at least, the precision of these ^{1}H NMR results is questionable.

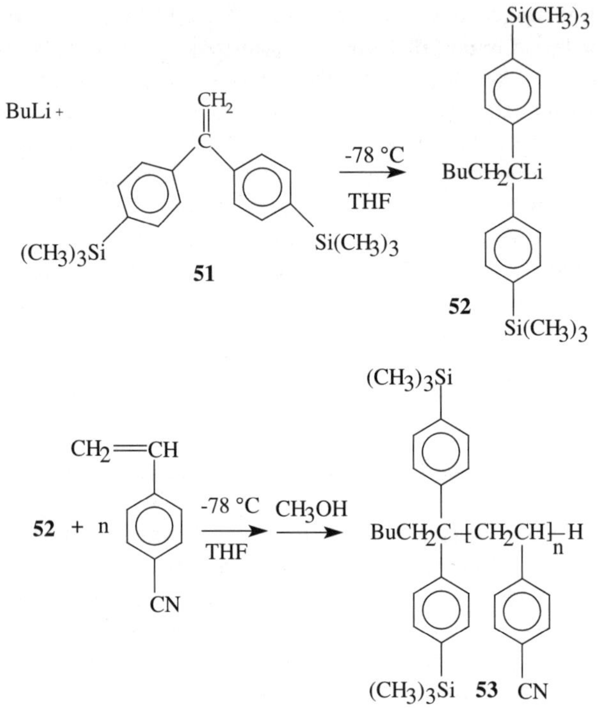

Scheme 22.

6.3.8
Fluorescent Group Labeling

Anionic functionalization methodology based on addition reactions to 1,1-diphenylethylenes can also be utilized for the preparation of polymers labeled with fluorescent groups by reactions with 1-phenyl-1-arylethylenes. Thus, poly(styryl)lithium was quantitatively labeled with a fluorescent naphthalene end group (see 56) via the reaction with 1-(2-naphthyl)-1-phenylethylene (55) as shown in Eq. (38) [174]:

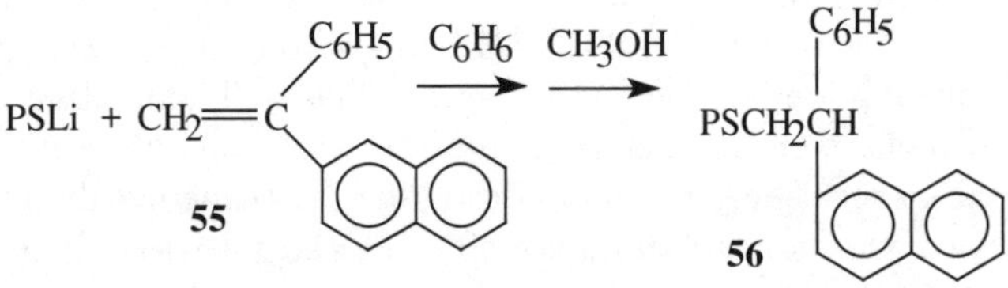

$$\text{(38)}$$

The addition reaction was completed in 8 h using a 0.1 molar excess of 55. The adduct of 1-(1-naphthyl)-1-phenylethylene with butyllithium (57) was used to initiate the anionic polymerization of methyl methacrylate at −78 °C in THF;

$$C_4H_9CH_2\underset{\underset{\textbf{57}}{|}}{\overset{\overset{C_6H_5}{|}}{C}}Li \text{ (1-naphthyl)}$$

the degree of labeling was reported to be 93% [175]; a more recent study concludes that essentially quantitative labeling can be obtained with this method [176]. The molecular weight distribution of the resulting PMMA was reported to be narrow ($M_w/M_n=1.07$), indicating that crossover to methyl methacrylate is competitive with propagation under these conditions. The initiator **57** has been used to prepare α-(1-naphthalene)-labeled poly(trimethylsilyl methacrylate) using analogous procedures [177].

This method has also been utilized for the preparation of pyrene end-labeled polymers [178] as shown in Eq. (39):

$$PLi + CH_2{=}\underset{\underset{\textbf{58}}{}}{\overset{\overset{C_6H_5}{|}}{C}}(pyrenyl) \xrightarrow{CH_3OH} PCH_2\overset{\overset{C_6H_5}{|}}{CH}(pyrenyl) \quad (39)$$

Spectroscopic analyses [λ_{max} (dioxane) 330, 346 nm; ^{1}H and ^{13}C NMR] indicated that these reactions proceed quantitatively to produce the polymers with pyrene fluorescent labels. The addition of poly(styryl)lithium with **58** (0.2 molar excess) was completed in approximately 1 h. In contrast, it was estimated that the addition of poly(butadienyl)lithium to **58** would require 1.5 months for completion at 25 °C. This is consistent with the unreactivity of poly(dienyl)lithiums with DPE as discussed previously in Sects 3.3 and 5.2. It was necessary to add a small amount of THF ([THF]/[Li]=12.3) to obtain a reasonable rate of addition; less than 2 h was required in the presence of THF [178].

The adduct of sec-butyllithium and a 0.2 molar excess of 1-(9-phenanthryl)-1-phenylethylene (**59**) has been used at –78 °C in THF to prepare α-phenanthrene-labeled poly(methyl methacrylates) [179]. Unfortunately, the initiator efficiency was only 80–88% for this initiator; consequently, the observed number average molecular weights were higher than the calculated values. Although the labeling efficiency calculated by UV-visible spectroscopy ranged from 0.90 to 0.99, the authors concluded that it was highly probable that all of the polymer chains were labeled with the fluorescent group [179].

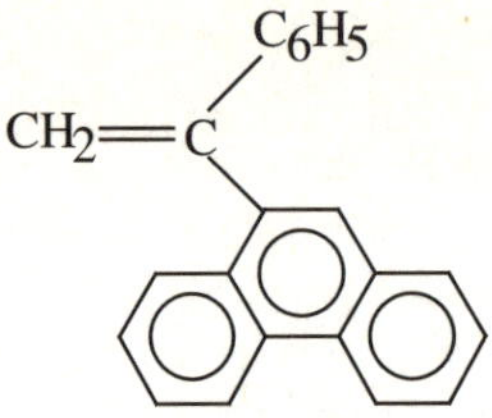

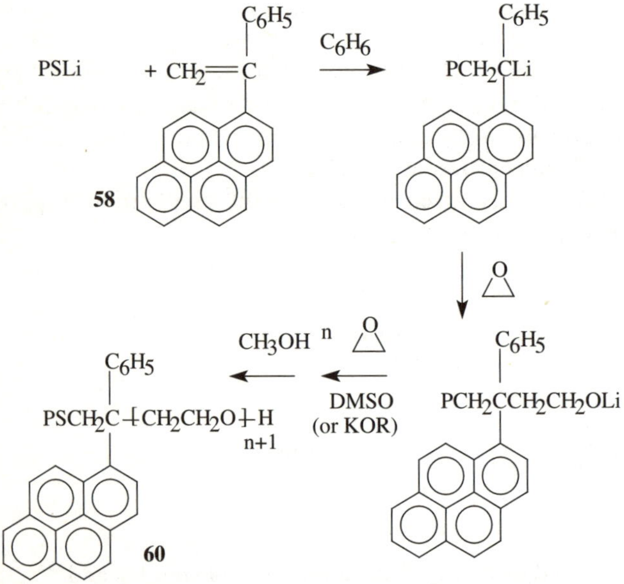

Scheme 23.

Since this fluorescent labeling methodology is a living functionalization reaction, the resulting living fluorescent-labeled polymers can be used to initiate the polymerization of a second monomer to produce a block copolymer with the label at the block interface as discussed previously. For example, this procedure has been used to prepare polystyrene-*block*-poly(ethylene oxide) copolymers with both pyrene (**60**) (see Scheme 23) and naphthalene fluorescent groups at the interface between the two blocks [180–182]. Lithium was used as the counterion to prepare well-defined, quantitatively-ethylene oxide-functionalized polystyrenes in benzene solution [183]. However, under these conditions, it is not possible to polymerize ethylene oxide [183]. Therefore, it was necessary to add either dimethylsulfoxide [180, 181] or a potassium alkoxide [182] to promote ethylene oxide block formation as shown in Scheme 23. These diblock copolymers were fractionated to obtain pure diblock copolymer

(M_w/M_n=1.05–1.09). The fluorescent labels were incorporated essentially quantitatively as determined by UV-visible spectroscopy, i.e., 98% for the pyrene-labeled diblock (**60**) and 99% for the naphthalene-labeled diblock.

Riess and coworkers [184, 185] have prepared analogous anthracene- and phenanthrene-labeled polystyrene-*block*-poly(ethylene oxide) polymers using cumylpotassium as initiator in THF. The styrene polymerization and fluorescent-group labeling were performed in THF at –78 °C, followed by addition of ethylene oxide and warming to room temperature. Unfortunately, this methodology yielded diblock copolymers with relatively broad molecular weight distributions (M_w/M_n=1.15–1.18). However, analysis of the fluorescent-group labeling efficiency by UV-visible spectroscopy indicated that essentially all of the copolymer chains were singly labeled.

The generality and efficiency of this fluorescent-labeling methodology has been demonstrated by preparation of polystyrene-*block*-poly(*tert*-butyl acrylate) [186] and polystyrene-*block*-poly(methyl methacrylate) [187] diblock copolymers with fluorescent groups at the block junctions.

6.3.9
Copolymerization with Substituted 1,1-Diphenylethylenes

The anionic copolymerization of substituted 1,1-diphenylethylene derivatives with copolymerizable monomers (M) will result in polymers with multiple functional groups along the polymer chain as illustrated in Eq. (40):

$$CH_2=C(C_6H_4N(CH_3)_2)_2 + nM \xrightarrow{BuLi} \text{polymer with } N(CH_3)_2 \text{ groups} \quad (40)$$

The number of functional groups per polymer molecule can be controlled by the monomer feed ratio and the molecular weight.

The anionic copolymerization of styrene and 1-(4-dimethylaminophenyl)-1-phenylethylene in benzene has been investigated [188]. As discussed previously in Sect. 5, Yuki and coworkers [125, 126, 129, 133–136] have developed the formalism for analyzing the kinetics of copolymerization of 1,1-diphenylethylene (M_2) with styrene and diene monomers (M_1). It was assumed that the 1,1-diphenylethylene derivative, M_2, does not add to itself due to steric effects, i.e., k_{22}=0, as discussed previously in Sect. 5. Thus, the monomer reactivity ratio for M_2 is zero, i.e., $r_2=k_{22}/k_{21}$=0. It was also assumed that the styrene monomer is completely consumed at the end of the polymerization

($[M_1]=0$) and that M_2 is in excess, i.e., there is still unreacted 1-(4-dimethylaminophenyl)-1-phenylethylene after the copolymerization. The resulting copolymerization equation is shown in Eq. (41) where r_1 is not unity and $[M_1]_0$, $[M_2]_0$ and $[M_1]$, $[M_2]$ are the initial and final monomer concentrations, respectively:

$$\ln \frac{[M_2]}{[M_2]_0} + \frac{1}{(r_1-1)} \ln \left[\frac{[M_1]_0}{[M_2]_0} (r_1-1) +1 \right] = 0 \tag{41}$$

The copolymerization of styrene with a 0.5 molar excess of 1-(4-dimethylaminophenyl)-1-phenylethylene produced a copolymer ($M_n=1.6 \times 10^4$) with 24 amine groups per chain [188]. Analogously, the copolymerization of styrene with a 2.6 molar excess of 1-(4-dimethylaminophenyl)-1-phenylethylene produced a copolymer ($M_n=3.8 \times 10^4$) with 37 amine groups per chain. These copolymers did not move on TLC plates in toluene, i.e., $R_f=0$, due to the large number of amine groups in the polymer molecules. The average value for r_1 was determined to be 5.6, which means that styrene is 5.6 times as reactive as the amine-substituted 1,1-diphenylethylene towards the poly(styryl)lithium anion. The corresponding value of r_1 is 0.4 for the copolymerization of styrene and 1,1-diphenylethylene in benzene at 30 °C [126]. These results are reasonable, since the addition reaction of 1,1-diphenylethylene derivatives to poly(styryl)lithium in benzene at room temperature has a Hammett ρ value of +1.8 [95]. The crossover reaction rate of poly(styryl)lithium with 1-(4-dimethylaminophenyl)-1-phenylethylene (k_{12}) is expected to be much less than that of poly(styryl)lithium to 1,1-diphenylethylene, due to the strong electron-donating effect of the dimethylamino group ($\sigma=-0.32$) [104]. Thus, the monomer reactivity ratio, r_1 ($r_1=k_{11}/k_{12}$), is expected to be larger for the copolymerization of styrene with 1-(4-dimethylaminophenyl)-1-phenylethylene than in the copolymerization of styrene and 1,1-diphenylethylene as observed.

It was anticipated that the copolymerization of substituted 1,1-diphenylethylenes with dienes such as butadiene and isoprene would be complicated by the very unfavorable monomer reactivity ratio for the addition of poly(dienyl)lithium compounds to 1,1-diphenylethylene [133, 134]. Yuki and Okamoto [133, 134] calculated values of $r_1=54$ and $r_1=29$ in hydrocarbon solutions for the copolymerization of 1,1-diphenylethylene (M_2) with butadiene (M_1) and isoprene (M_1), respectively. Although the corresponding values in THF are r_1(butadiene)=0.13 and r_1(isoprene)=0.12, this would not be an acceptable solution since THF is known to form polymers with high 1,2-microstructures [3]. Anionic copolymerizations of butadiene (M_1) with excess 1-(4-dimethylamino-phenyl)-1-phenylethylene (M_2) were conducted in benzene at room temperature for 24–48 h using sec-butyllithium as initiator [189]. Anisole, triethylamine and tert-butyl methyl ether were added in ratios of [B]/[RLi]=60, 20, 30, respectively, to promote copolymerization and minimize 1,2-enchainment in the polybutadiene units. Narrow molecular weight distribution copolymers with $M_n=14 \times 10^3$ to 32×10^3 ($M_w/M_n=1.02-1.03$) and 8, 12, and 30 amine groups per chain for anisole, triethylamine, and *tert*-butyl methyl ether, respectively, were obtained. The butadiene monomer reactivity ratios (r_1) and microstructures (% of 1,2-) were 42 (14%), 33 (29%), and 14 (58%) for anisole, triethylamine, and *tert*-butyl methyl ether, respectively. It was anticipated that the addition of alkali metal (Mt) salts, e.g., alkoxides, at low levels ([MtOR]/[Li]<<1) would promote copolymerization without significant increases in vinyl microstructure [3, 190, 191]. Copolymerizations of 1-(4-dimethylaminophenyl)-1-phenylethylene (M_2) with butadiene in the presence of potassium *tert*-amyloxide ([KOR]/[Li]=0.025) readily incorporated amine groups into the polybutadiene copolymer (r_1=1.35) and the amount of 1,2-microstructure was only 28% [189].

6.3.10
Conclusions Regarding 1,1-Diphenylethylene Functionalizations

The addition reactions of simple and polymeric organolithium compounds with substituted 1,1-diarylethylenes provide a general method for the synthesis of functionalized and labeled polymers. With this method in conjunction with appropriate protecting groups and reaction conditions, a wide variety of well-characterized, quantitatively functionalized, and labeled polymers and copolymers can now be prepared with diverse molecular structures. Polymers can be prepared with functional groups at the initiating chain end, at the terminating chain end, within the polymer chain by copolymerization, and at the junction between blocks.

7
Preparation of 1,1-Diphenylethylene-Functionalized Macromonomers and their Applications for Synthesis of Heteroarm, Star-Branched Polymers

Macromonomers are linear macromolecules carrying some polymerizable functional group at their chain end; the polymerizable functional groups can be at one chain end or at both chain ends [149–151, 192–194]. The important feature of macromonomers is that they can undergo copolymerization with other monomers by a variety of mechanisms to form comb-type, graft copolymers as shown in Eq. (42):

$$y \text{ \textasciitilde\textasciitilde\textasciitilde} - X + mM_2 \longrightarrow \left[-[M_2]_n - X - M_2 - \right]_z \quad (42)$$

macromonomer

m >> y

comb-type graft copolymer

Generally the polymerizable functional groups at the chain end of macromonomers are vinyl groups (**61**).

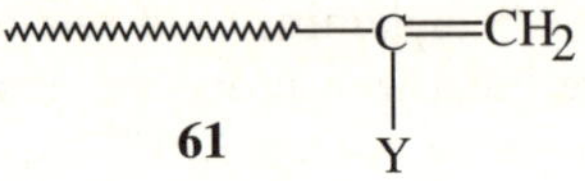

〜〜〜〜〜〜〜〜〜—C=CH$_2$
         61 |
               Y

Macromonomers provide a unique method of preparing graft copolymers with control of the branch structure. If a well-defined macromonomer can be prepared with a low degree of compositional heterogeneity, then copolymerization of this macromonomer with other monomers will form a graft copolymer in which the structure of the branches is also well defined. However, the incorporation of the macromonomer will be controlled by the statistics of the copolymerization process; thus, there will be compositional heterogeneity associated with the number of graft branches per molecule and the distribution of the graft branches along the polymer backbone in the graft copolymer.

In principle, living anionic polymerization with a non-homopolymerizable macromonomer can provide a method of preparing branched and graft polymers with control of all of the structural parameters and with low degrees of compositional heterogeneity as illustrated in Scheme 24 [3, 195]. Thus, living

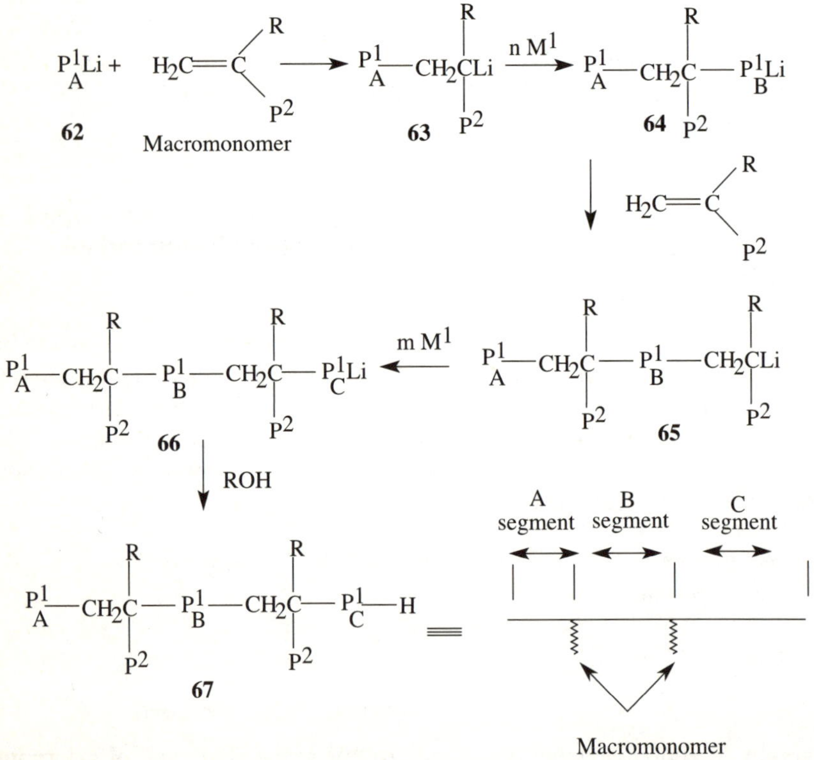

Scheme 24.

anionic polymerization will form the first (A) backbone segment, 62, which will react with a non-homopolymerizable macromonomer to add only one macromonomer unit to the chain end to form 63 and maintain the living nature of the polymerization. Addition of more backbone-forming monomers (nM^1) will then generate a new (B) backbone segment whose length will be defined by the ratio of the grams of monomer added to the moles of active chain end of 63. The living anionic polymer 64 can then react with another equivalent of non-homopolymerizable macromonomer to place another well-defined graft branch at the precise point on the backbone defined by the B segment length and form 65. Addition of more backbone-forming monomer (mM^1) to the adduct 65 will then generate a new (C) backbone segment whose length will be defined by the ratio of the grams of monomer added to the mole of active chain end 65. At this point the whole sequence of macromonomer followed by comonomer addition could be repeated to generate more graft branches at specific locations along the backbone, or the living polymer (66) could be terminated to form the precisely-defined graft copolymer (67) with two grafted branches located at segment distances of A and A+B along the polymer backbone.

It was deduced by several research groups that a macromonomer with a 1,1-diphenylethylene functional group could function as a useful non-homopolymerizable macromonomer for synthesis of well-defined, heteroarm star and graft copolymers. The first 1,1-diphenylethylene macromonomer synthesis was reported by Isono and coworkers [196]. This macromonomer was used to synthesize a hetero, three-armed star-branched polymer (72) with polydimethylsiloxane, polystyrene and poly(*tert*-butyl methacrylate) arms as shown in Scheme 25. The 1,1-diphenylethylene-functionalized polydimethylsiloxane (PDMS) macromonomer, 70, was prepared by polymerization of hexamethyltrisiloxane (D_3) using the lithium salt of 69 as initiator. It was assumed that 1,1-diphenylhexyllithium reacts only with the silanol group in 69 and does not add to the 1,1-diphenylethylene group. This assumption was supported by the ^1H NMR resonance spectrum of 70 that exhibited vinylidene protons at δ 5.48 ppm which when integrated relative to the methyl protons in the dimethylsiloxane repeating units provided a number average molecular weight which was in good agreement with the value determined by membrane osmometry. Unfortunately, the molecular weight distributions for the DPE-functionalized PDMS macromonomers were somewhat broad ($M_w/M_n=1.2_9$–1.3_6). For the reaction of poly(styryl)lithium with 70, it was assumed that PSLi reacts only by addition to the 1,1-diphenylethylene unit and does not add to the polydimethylsiloxane units. However, there is ample precedent for the reactivity of organolithium compounds with alkoxysilanes [19] and the assumption that these reactions do not occur is a severe fault of this procedure. It would be expected that PSLi will attack the siloxane bonds in 70, which could cause cleavage of the PDMS chain as well as multiple linking of PSLi with the macromonomer. For the final step in this synthesis, the living adduct, 71, was cooled in THF and used to initiate the polymerization of *tert*-butyl methacry-

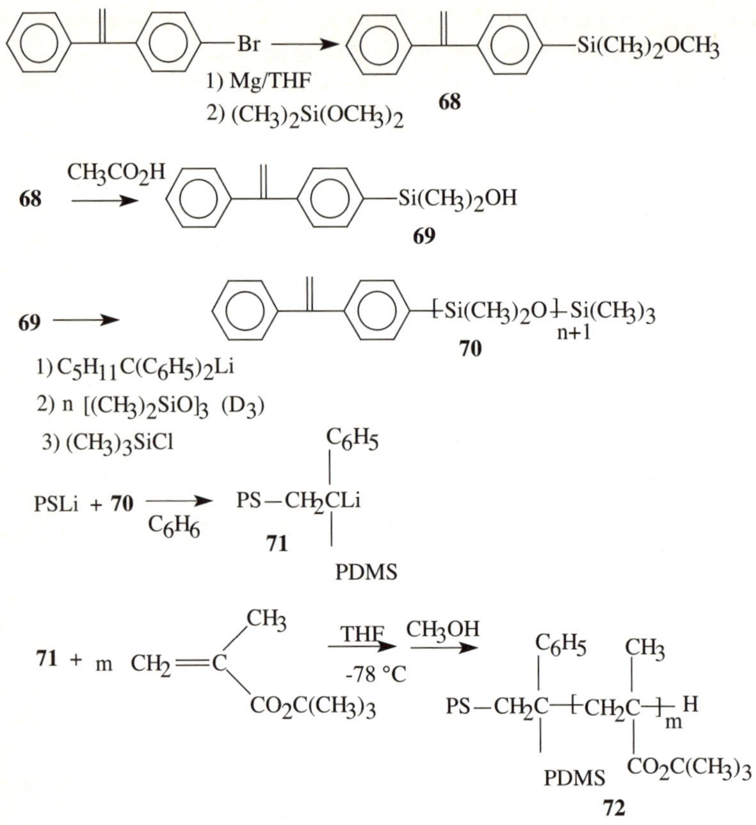

Scheme 25.

late to form the hetero, three-armed star-branched polymer, **72**. An excess of macromonomer was used and the resulting star-branched polymer, **72**, exhibited broad, multimodal molecular weight distributions. Fractionation provided samples with "fairly narrow molecular weight distributions"; however, no specific characterization data other than relative ^{1}H NMR signal intensities and the overall number average molecular weight (osmometry) was provided.

The second reported synthesis and application of 1,1-diphenylethylene macromonomers provided less ambiguous results and illustrated the potential of these macromonomers in synthesis of well-defined, hetero, three-armed, star-branched polymers [197, 198]. 1,1-Diphenylethylene-functionalized polystyrene macromonomers (**74, 76**) were prepared by the addition of poly(styryl)lithium with either 1,3-bis(1-phenylethenyl)benzene (**73**, MDDPE) [198] or 1,4-bis(1-phenylethenyl)benzene (**75**, PDDPE) [197] as shown in Eqs. (43) and (44):

Applications of 1,1-Diphenylethylene Chemistry in Anionic Synthesis... 127

(43) PSLi + **73** MDDPE →[1) -78 °C, THF/C₆H₆][2) CH₃OH] **74**

(44) PSLi + **75** PDDPE →[1) C₆H₆][2) CH₃OH] **76**

These syntheses were based on the observation fact that, although both MDDPE and PDDPE react rapidly with two moles of simple or polymeric organolithium compounds in hydrocarbon solution to form the corresponding dilithium adducts [3, 89,199–203], in THF monoaddition is reported. For example, McGrath and coworkers [204] reported that the monoadduct of MDDPE and sec-butyllithium could be prepared in THF at −78 °C. Similar observations have been confirmed by Ma [205] and Ikker and Möller [206].

The functionalities of these macromonomers could be determined by both UV-visible spectroscopic analysis (λ_{max} for DPE at 260 nm) and by ^{1}H NMR spectroscopy (vinylidene protons at δ 5.4 ppm). The functionality of the macromonomer **74** (M_n=3400) was estimated to be 0.6 by UV analysis and 0.83 by NMR; no dimer was observed by SEC analysis [198]. The stoichiometric addition of PSLi (M_n=16,000) occurred quantitatively with no residual macromonomer observed by SEC [198].

It was found that the addition of PSLi with PDDPE (**75**) exhibited less tendency to form the corresponding diadduct in both hydrocarbon solution and in the presence of THF [197]. Presumably dimer formation is less favorable in the case of the *para*-substituted PDDPE compared to the *meta*-substituted MDDPE because the negative charge in the monoadduct can be delocalized into all three aromatic rings and the remaining vinyl group in PDDPE but not in MDDPE. For example, even with a fourfold excess of MDDPE, the addition of PSLi in benzene at room temperature produced the dimer in 87% yield [197]. In contrast, with only a twofold excess of PDDPE, the reaction with PSLi produced only 7% dimer [197]. At a temperature of 5–8 °C in benzene in the presence of THF ([THF]/[Li]=20), the 1,1-DPE-functionalized polystyrene macromonomer from PSLi (M_n=5400) and **75** was obtained with a functionality of 98% (^{1}H NMR and UV).

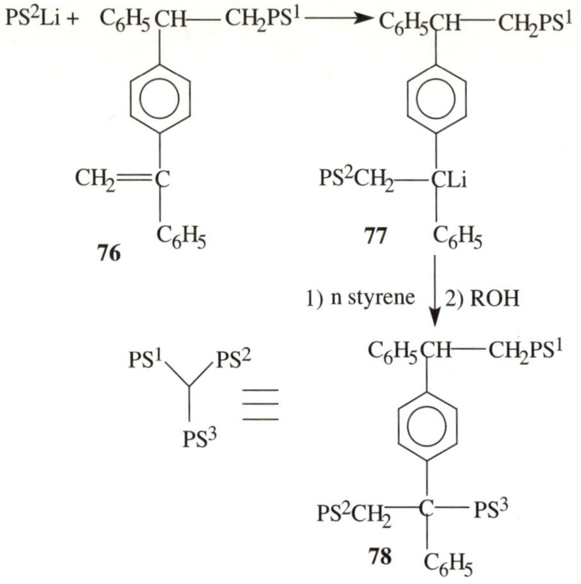

Scheme 26.

For the synthesis of three-armed, heteroarm, star-branched polymers (see Scheme 26), the first step involves the addition of a polymeric organolithium compound, e.g., poly(styryl)lithium, with the 1,1-diphenylethylene-functionalized polystyrene macromonomer, **76**, to form the corresponding coupled product, **77**, a polymeric 1,1-diphenylalkyllithium [197, 207]. For stoichiometric amounts of poly(styryl)lithium (M_n=15,300) and macromonomer (M_n=5400), it was found that the efficiency of this coupling reaction is >96%. This result also shows that the vinyl functionality of the macromonomer is >96%. Finally, the third arm [M_n(calc)=30,000] was formed by addition of monomer, e.g., styrene, in the presence of THF ([THF]/[Li]=20), to promote the crossover reaction; this reaction was completed in 1.5 h. SEC analyses of the heteroarm star polymer product (**78**) showed that each of these steps proceeded efficiently to give the expected products. Only relatively small amounts of non-star products were observed by SEC; peaks were observed whose retention volumes corresponded to a small amount of unreacted macromonomer (**76**) and a small amount of polystyrene corresponding to the second arm (PS²Li). A narrow molecular weight distribution star product (M_w/M_n=1.02) was easily obtained by one fractionation step [197, 207]. The observed molecular weight [M_n(universal calibration)=55,400; M_n(membrane osmometry)=58,300]] was somewhat larger than the calculated molecular weight [M_n(calc)=50,700].

These procedures have also been applied to the synthesis of 1,1-diphenylethylene-functionalized polybutadiene macromonomers (**79**) [208, 209]. The general reaction scheme is shown in Eq. (45):

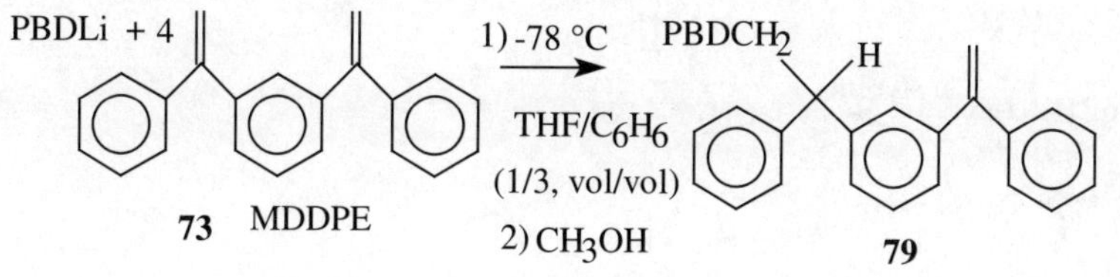

[Scheme at top showing reaction (45): PBDLi + 4 MDDPE (73) → via 1) -78 °C, THF/C$_6$H$_6$ (1/3, vol/vol), 2) CH$_3$OH → PBDCH$_2$ product (79)]

It was found that it was necessary to utilize a fourfold excess of MDDPE (**73**) at −78 °C in order to obtain a polybutadiene macromonomer (**79**) with high functionality (98–99% by UV spectroscopy) [209]. Under these conditions (3.5 h reaction time), the amount of dimer formation ranged from 1% to 2.8% as determined by SEC.

An alternative procedure has been utilized to prepare 1,1-diphenylethylene-functionalized poly(ethylene oxide) macromonomers as shown in Scheme 27 [210, 211]. This macromonomer (**81**) reacted quantitatively with poly(styryl)lithium to form the corresponding living diblock copolymer adduct. After cooling to 5 °C, *tert*-butyl methacrylate was added and polymerized for ca. 2 h. After quenching with acidic methanol, the hetero, three-armed, star-branched, ABC-type block copolymer was isolated. The molecular weight determined by SEC (universal calibration; M_n=19,500; M_w/M_n=1.14) and by ^{1}H NMR (M_n=17,300) were somewhat higher than the calculated value (M_n=15,500) [211].

Using the silylamine-functionalized 1,1-diphenylalkyllithium initiator, **39**, to polymerize styrene, the corresponding functionalized poly(styryl)lithium, **40** (see Scheme 18), was reacted with MDDPE to form the silyl-amine-functionalized, 1,1-diphenylethylene-functionalized polystyrene macromonomer (**82**) as shown in Scheme 28 [212]. Less than 2% dimer formation was observed by SEC.

Stadler and coworkers [213] have described another method for synthesis of 1,1-diphenylethylene-functionalized polystyrene macromonomers (**84**) as

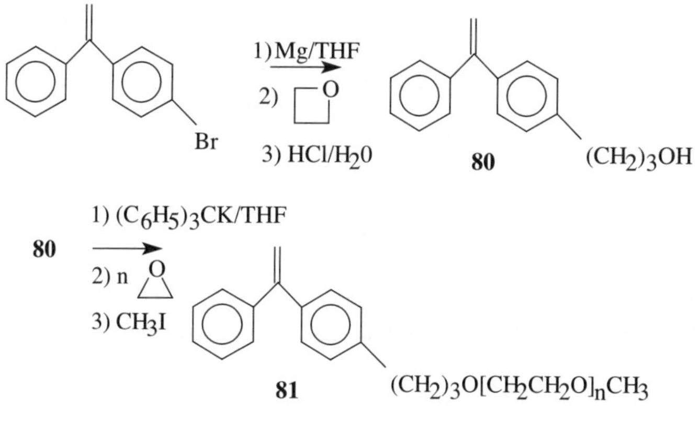

[Scheme 27 reactions:
- Ar-Br compound → 1) Mg/THF, 2) propylene oxide, 3) HCl/H$_2$O → **80** with (CH$_2$)$_3$OH substituent
- **80** → 1) (C$_6$H$_5$)$_3$CK/THF, 2) n ethylene oxide, 3) CH$_3$I → **81** with (CH$_2$)$_3$O[CH$_2$CH$_2$O]$_n$CH$_3$ substituent]

Scheme 27.

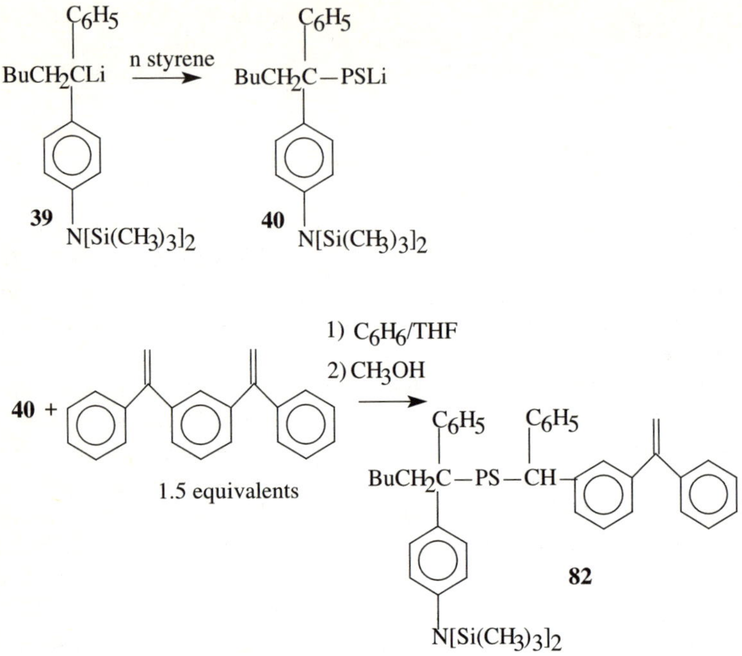

Scheme 28.

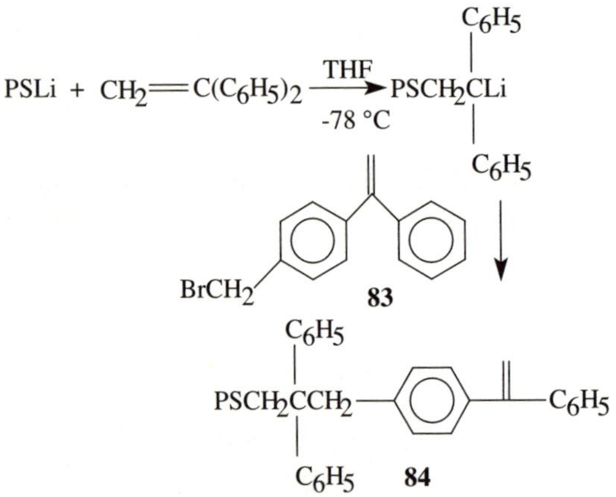

Scheme 29.

shown in Scheme 29. Poly(styryl)lithium prepared in THF at −78 °C was first end-capped with 1,1-diphenylethylene to reduce the reactivity and increase the steric hindrance of the chain end to minimize addition to the 1,1-diphenylethylene group in **83** and also to minimize lithium-halogen exchange [19]. The reaction of end-capped anion with excess **83** was rapid. The resulting 1,1-

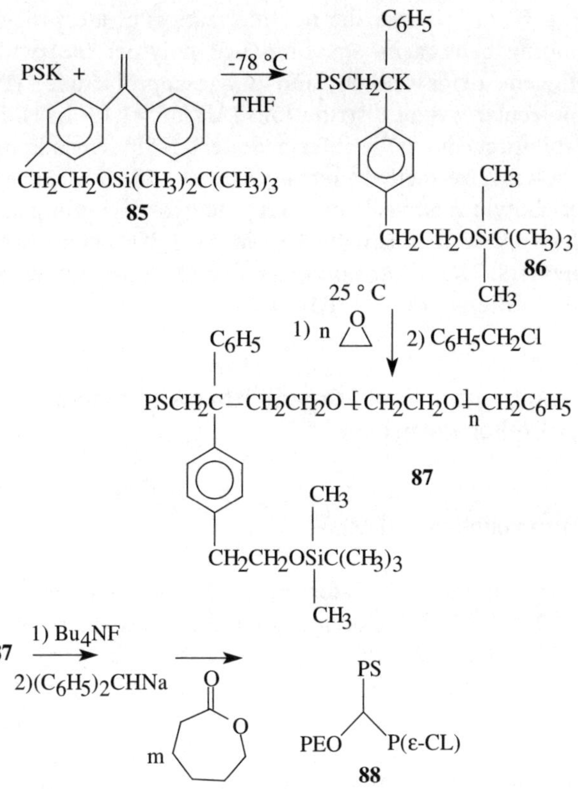

Scheme 30.

diphenylethylene-functionalized polystyrene macromonomer, **84**, exhibited a narrow molecular weight distribution ($M_w/M_n=1.03$). The functionality was reported to be quantitative by UV spectroscopic analysis. Poly(butadienyl)-lithium was prepared in THF at $-10\,°C$ and reacted with **84** to form the corresponding living diblock copolymer which was then used to initiate the polymerization of methyl methacrylate at $-69\,°C$. The final product was contaminated with significant amounts of polystyrene, polybutadiene, the diblock, and other unidentified impurities. After extraction, a very broad peak with a long low molecular weight tail was obtained ($M_w/M_n=1.13$).

Riess and coworkers [214, 215] have also developed a unique method to synthesize heteroarm, ABC-type, three-armed, star-branched block copolymers as shown in Scheme 30. Poly(styryl)potassium was reacted with a *tert*-butyldimethylsiloxy-protected, hydroxyethylated 1,1-diphenylethylene (**85**) to form a siloxy-functionalized polymeric 1,1-diphenylalkylpotassium (**86**), that was then used to initiate polymerization of ethylene oxide as shown in Scheme 30 [214]. Samples of the diblock copolymers, **87**, exhibited rather broad molecular weight distributions ($M_w/M_n=1.15-1.16$). After removal of the protecting group with tetrabutylammonium fluoride the corresponding alkoxide was used to initiated the polymerization of ε-caprolactone to form the corresponding heteroarm, star-branched polymer (**88**) with polystyrene (PS), poly(ethylene oxide) (PEO), and poly(ε-caprolactone) [P(ε-CL)] arms with broad molecular weight distributions (M_w/M_n=1.19–1.41). In another application of this procedure, the intermediate 1,1-diphenylalkylpotassium, **86**, was used to polymerize methyl methacrylate at –78 °C [215]. After deprotection of the *tert*-butyldimethylsiloxy group, the hydroxyl group as used to polymerize ethylene oxide to form the PS-PMMA-PEO, hetero, three-arm, star-branched polymers. These star polymers also exhibited rather broad molecular weight distributions (M_w/M_n=1.16–1.26).

8
Difunctional and Trifunctional Organolithium Initiators Based on 1,1-Diphenylethylene

8.1
Difunctional Organolithium Initiators

Difunctional organolithium initiators are of considerable interest for the preparation of A-B-A triblock copolymers, telechelic polymers, and macrocyclic polymers [3, 64]. Although triblock copolymers can be prepared with monofunctional initiatiors using a three-step, sequential monomer addition process, with difunctional initiators they can be formed in a more efficient two-step process. Difunctional initiators also provide a methodology to prepare new triblock copolymers that cannot be prepared by the three-step, sequential monomer addition route because the chain ends formed from the first monomer (forming block A) are too stable to initiate the polymerization of the second monomer (forming block B). For example, difunctional initiators have been used for the direct synthesis of poly(methyl methacrylate)-*block*-polybutadiene-*block*-poly(methyl methacrylate) [216] and poly(*tert*-butyl methacrylate)-*block*-polyisoprene-*block*-poly(*tert*-butyl methacrylate) [217] that cannot be prepared by sequential monomer addition. Difunctional initiators provide direct, efficient methods for the formation of α,ω-difunctional polymers, i.e., telechelic polymers, by termination reactions of the polymeric α,ω-dianions with electrophilic functionalization agents [2]. Analogously, termination of the α,ω-dianions with a difunctional, electrophilic coupling agent under high dilution conditions promotes intramolecular cyclization reactions to form macrocyclic polymers [218].

Aromatic radical anions, such as lithium dihydronaphthylide, are efficient difunctional initiators [6]. However, the necessity of using polar solvents for their formation and use limits their utility for diene polymerization since the unique ability of lithium to provide high 1,4-polydiene microstructure is lost in polar media [1–3]. The strategies which have been employed in the search for hydrocarbon-soluble, dilithium initiators are generally based on the reaction of an aromatic divinyl precursor with two moles of butyllithium [64, 89]. Because of the tendency of organolithium chain ends to exist predominantly in hydrocarbon solution as associated, electron-deficient species such as dimers, tetramers, and hexamers [3, 8, 19], attempts to prepare dilithium initiators in hydrocarbon solutions have generally resulted in formation of insoluble, three-dimensionally associated species [89]. These precipitates are not effective initiators because their heterogeneous initiation reactions with monomers tend to result in broader molecular weight distributions (M_w/M_n>1.1), at best. Soluble analogs of these difunctional initiators have been prepared by addition of small amounts of weakly basic additives such as triethylamine or anisole, that have relatively minor effects on diene microstructure [49, 219]. Another method to solubilize these initiators is to use a seeding technique, whereby small amounts of diene monomer are added to form a hydrocarbon-soluble, oligomeric dilithium initiating species [49, 220–222]. In general, none of these methods have provided a simple, stable, reproducible, hydrocarbon-soluble dilithium initiator.

8.1.1
1,4-Dilithio-1,1,4,4-tetraphenylbutane

Success in designing useful, hydrocarbon-soluble, dilithium initiators can be traced to the pioneering studies of Morton and Fetters [49]. They prepared a useful dilithium initiator by the dimerization of 1,1-diphenylethylene with lithium in cyclohexane in the presence of anisole (15 vol. %). Although the initiator was soluble in this mixture, it precipitated from solution when added to the polymerization solvent (cyclohexane or benzene). The polystyrene-*block*-polyisoprene-*block*-polystyrene and poly(α-methylstyrene)-*block*-polyisoprene-*block*-poly(α-methylstyrene) triblock copolymers which were formed by sequential monomer addition using this initiator contained polyisoprene center blocks with >90% 1,4-microstructure (65–70% *cis*-1,4; 20–25% *trans*-1,4). Furthermore, the polymers had predictable molecular weights and narrow molecular weight distributions (M_w/M_n=1.07). Thus, contrary to expectations based on the increased stability of 1,1-diphenyl-alkyllithiums compared to poly(isoprenyl)lithium based on the pK_a values of the corresponding conjugate acids [3, 43], the dilithium initiator was an efficient initiator for isoprene polymerization [M_n(calc)≈M_n(obs)] and also initiated isoprene polymerization at a rate which was competitive with propagation since the molecular weight distribution was narrow (M_w/M_n=1.07). Until this report by Morton and Fetters [49], 1,1-diphenylalkyllithiums had only been used to initiate polymerization of methyl methacrylate [46–48], which forms a more stable ester enolate anion [43].

8.1.2
1,3-Bis(1-phenylethenyl)benzene

Schulz and Höcker [223] first reported that the reaction of *sec*-butyllithium with 1,3-bis(1-phenylethenyl)benzene (MDDPE, 73) forms a hydrocarbon soluble dilithium initiator in aromatic solvent. Leitz and Höcker [199] first reported kinetic studies of the reaction of *sec*-butyllithium with 1,3-bis(1-phenylethenyl)benzene (MDDPE, **73**) and 1,4-bis(1-phenylethenyl)benzene (PDDPE, **75**). For PDDPE, the rate of the first addition of *sec*-butyllithium in toluene ($k_1 = 20.6$ l/mol s) is much faster than the second addition in toluene ($k_2 = 1.51$ l/mol s) (Eq. 46):

$$ \text{(46)} $$

For **73**, the rate of the second addition in toluene ($k_4 = 19.8$ l/mol s) was equal to the rate of the first addition (Eq. 47),

$$ \text{(47)} $$

and the same as the rate of the first addition for **75**. McGrath and coworkers [204] reported that the rate of the second addition of *sec*-butyllithium to **73** is actually an order of magnitude faster than that of the first addition in cyclohexane.

These unexpected results can be explained by the expectation that the negative charge in the monoadduct, **87**, can be conjugated with the *para*-1-phenylethenyl group as shown in Eq. (46). This would tend to make this doublebond less reactive toward addition of another negatively charged organolithium. However, the negative charge in **89** is not delocalized into the *meta*-1-phenylethenyl group. Also, in hydrocarbon solution in the absence of extended conjugation in **89**, the small positive lithium counterion would tend to localize negative charge in the monoanion, **89**, in the vicinity of lithium at the 1,1-diphenyl-substituted carbon. Another unexpected aspect of the addition of alkyllithiums to 1,3-bis(1-phenylethenyl)benzene is the fact that although the dilithium initiator, **90**, is soluble in hydrocarbon media such as cyclohexane, benzene and toluene (even at –20 °C), it is not soluble in *n*-hexane [200]. It is noteworthy that initiator **88**, like many other dilithium initiators, is not soluble in benzene and polymers prepared using this initiator exhibited broad molecular weight distributions (M_w/M_n=1.2–1.4) [220]. Other double diphenylethylene precursors for dilithium initiators and their solubilities are shown in Scheme 31 [89, 220, 224].

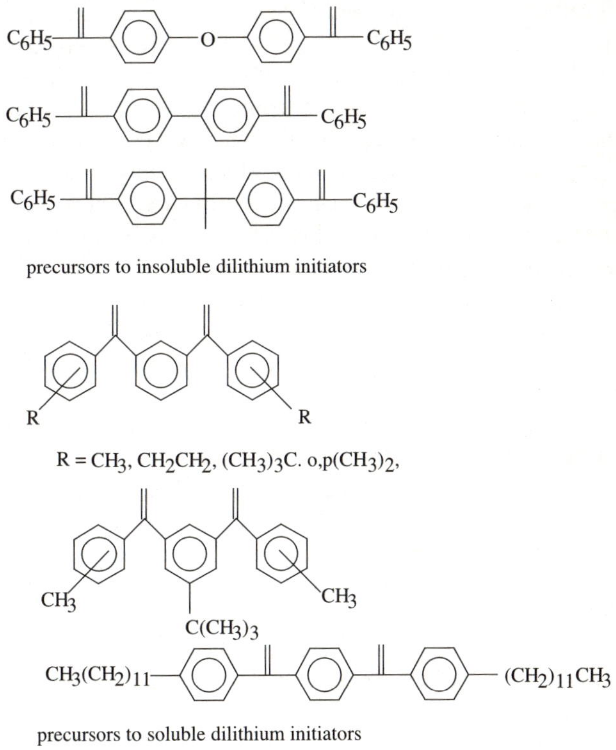

precursors to insoluble dilithium initiators

R = CH$_3$, CH$_2$CH$_2$, (CH$_3$)$_3$C. o,p(CH$_3$)$_2$,

precursors to soluble dilithium initiators

Scheme 31.

Tung and Lo [225] reported that dilithium initiator, **90**, functions as an efficient difunctional initiator for the preparation of homopolymers and triblock copolymers with relatively narrow molecular weight distributions. However, these workers provided few experimental details at that time. More recent reports have elaborated on the use of this initiator [200].

McGrath and coworkers [217] demonstrated that the use of the dilithium initiator, **90**, provides a very versatile method with respect to the ability to vary the chemical composition of the end blocks after first polymerizing a diene such as butadiene or isoprene. The α,ω-dilithiumpolydiene center block can be used to initiate the polymerization of polar monomers to form both end blocks simultaneously. This type of triblock copolymer with polar end blocks cannot be prepared by other block copolymer synthesis procedures. Thus, they prepared poly(*tert*-butyl methacrylate)-*block*-polyisoprene-*block*-poly(*tert*-butyl methacrylate) triblock copolymers. However, the resulting triblock copolymers did not exhibit well-defined structures. The molecular weight distributions tended to be broad (M_w/M_n=1.10–1.25) or bimodal.

A careful reexamination of the use of the initiator, **90**, formed by addition of 2 moles of *sec*-butyllithium with 1,3-bis(1-phenylethenyl)benzene in benzene showed that, although this dilithium initiator is soluble in hydrocarbon media, it produces polystyrenes and polybutadienes with multimodal molecular weight distributions and no control of molecular weight [88], in spite of reports to the contrary [200, 225]. Although later work erroneously attributed these multimodal distributions to inexact stoichiometry [226], association effects were clearly the cause as deduced from the results of identical experiments effected in the presence and in the absence of the Lewis base, THF [88]. In the absence of Lewis base, bimodal molecular weight distributions were obtained. However, using the same batches of initiator in the presence of stoichiometric amounts of THF ([THF]/[Li]=32–47), monomodal, narrow molecular weight distribution polystyrenes and polybutadienes were formed with number average molecular weights in good agreement with the values calculated from the ratio of grams of monomer divided by moles of initiator [88]. However, the polybutadiene microstructure corresponded to 42% 1,2, 31% *trans*-1,4, and 28% *cis*-1,4.

These results stand in sharp contrast to the results reported earlier by McGrath and coworkers [217], the work of Tung and Lo [225], and the efficiency and reactivity of other 1,1-diphenylalkyllithium initiators [49]. However they are in accord with unpublished data reported by Bywater [1]. Because of these contradictions, the experimental procedures utilized in these studies were scrutinized, since the bimodal distributions were obtained using high purity reagents and carefully controlled, high vacuum conditions [88].

Bimodal molecular weight distributions could result from either chain-end association effects or from inexact stoichiometry. Inexact stoichiometry could result in branching if excess MDDPE is used, or both monofunctional and difunctional growing chains if excess *sec*-butyllithium is used. For example, Bastelberger and Höcker [227] reported that, for the dilithium initiator, **91**,

(Eq. 48), bimodal molecular weight distributions result when the initiator solution contains both monofunctional and difunctional initiating species:

$$\text{C}_6\text{H}_5\text{-C(=CH}_2\text{)-C}_6\text{H}_4\text{-(CH}_2\text{)}_{10}\text{-C}_6\text{H}_4\text{-C(=CH}_2\text{)-C}_6\text{H}_5$$

$$\downarrow 2\ sec\text{-butyllithium}$$

$$\text{C}_4\text{H}_9\text{CH}_2\text{-C(Li)(C}_6\text{H}_5\text{)-C}_6\text{H}_4\text{-(CH}_2\text{)}_{10}\text{-C}_6\text{H}_4\text{-C(Li)(C}_6\text{H}_5\text{)-CH}_2\text{C}_4\text{H}_9$$

91

(48)

The effect of added Lewis base, THF, was investigated to distinguish inexact stoichiometry effects from chain-end association effects. It has been reported that Lewis bases such as THF promote the dissociation of aggregates of both simple and polymeric organolithium compounds to form either unassociated species or aggregates with a lower average degree of association depending on the organolithium compound, the nature of the base, and its concentration [3, 8, 19, 36, 113]. The addition of THF in amounts expected to be sufficient to promote dissociation of the diphenylalkyllithium aggregates ([THF]/[Li]=14–32) [113] produced monomodal molecular weight distribution polybutadienes and polystyrenes with M_n=26,000 and 10,000, respectively [88]. These observations and the fact that monomodal distributions result when higher molecular weight polymers are prepared to rule out effectively the possibility that the bimodal effects are due to inexact stoichiometry. These results were confirmed recently by Hadjichristidis and coworkers [228] for the polymerization of styrene using **90** in the presence of THF.

Although the addition of the Lewis base, THF, was effective in eliminating the bimodality obtained for both polybutadiene and polystyrene using **90** as initiator, this is not a generally satisfactory solution since it is known that diene microstructure is changed from high 1,4-enchainment to high 1,2-microstructure for organolithium-initiated diene polymerization in the presence of THF [1–3]. A solution to this conundrum was suggested after careful consideration of the experimental details for those literature reports dealing with the use of **90**, which reported monomodal molecular weight distributions, since no Lewis base additives were utilized. In general, these procedures utilized less than rigorous, i.e., not high vacuum, techniques; for example, some procedures describe the use of excess butyllithium, i.e., more than the stoichiometric amount required to generate the initiator [200]. The excess butyllithium was used to scavenge impurities in the solvent and monomer feed; otherwise some of the initiator itself would be consumed in scavenging impurities. This suggested that monomodal molecular weight distributions may have been obtained in the presence of added lithium alkoxide, which could have been formed by the reaction of the excess butyllithium with oxygen and/or hydroxylic impurities [19]. This hypothesis was tested by addition of

$$\underset{\substack{|\\CH_3}}{CH_3CH_2CHCH_2CLi(ROLi)_n} \overset{CH_3 \quad C_6H_5}{\underset{|\quad\quad|}{}}$$

⟨○⟩

$$CH_3CH_2CHCH_2CLi(ROLi)_n$$
$$\underset{CH_3 \quad C_6H_5}{|\quad\quad|}$$
92

93
(structure showing C$_6$H$_5$ and C$_5$H$_{11}$ substituted carbons bridged by Li atoms around two phenylene rings)

lithium *sec*-butoxide to the dilithium initiator, **90** ([LiOBu]/[**90**]=1/1). It was observed that the use of alkoxide-modified initiator produced low molecular weight (M_n=25,000), monomodal, narrow molecular weight distribution (M_w/M_n=1.1) polybutadienes [88]. These observed effects of added lithium alkoxide are consistent with earlier reports that the addition of these salts accelerates initiation reaction rates and depresses propagation rates for isoprene polymerizations using alkyllithium initiators [229, 230]. It is known that lithium alkoxide salts cross-associate with organolithium compounds and form a statistical distribution of mixed aggregates [231–237]. For example, equimolar amounts of *n*-butyllithium and lithium *tert*-butoxide form a tetrameric 1:1 complex in benzene solution [235, 236]. Thus, species such as **92** would be expected to be formed from mixtures of **90** and lithium alkoxides. These considerations suggest either that the mixed aggregate **92** is an effective initiator, or that preferential dissociation of diphenylalkyllithium species occurs from the mixed aggregates relative to the dissociation of the poly(butadienyl)lithium formed after the crossover reaction.

It is important to consider the possible reasons for the association effects which lead to bimodal molecular weight distributions for polymers formed using **90** as initiator in the absence of added Lewis base or lithium alkoxide. Leitz and Höcker [199] proposed that double diphenylethylene-based dilithium initiators form dimeric dianion aggregates (**93**) and that is why they are soluble in hydrocarbon solutions compared to other dilithium species. This type of dimeric structure is consistent also with the dimeric association of 1,1-diphenylalkyllithiums [36]. Szwarc and coworkers [238] have reported that analogous dilithium species based on 1,1-diphenylethylene units are dimeric in benzene solution.

The fact that residual UV-visible absorption at 438 nm due to unreacted diphenylalkyllithium initiator residues was observed when bimodal molecular weight distributions were obtained for initiator **90** [88] suggests that initiation reactions are slow relative to propagation reactions after one of the 1,1-diphenylalkyllithium initiator sites has crossed over to monomer.

Teyssie and coworkers [216] have reported that the analogous initiator formed by addition of 2 moles of *tert*-butyllithium with **73**, when first used to polymerize butadiene followed by low temperature polymerization of methyl methacrylate, forms poly(methyl methacrylate) end blocks that are bimodal. However, the sample used for testing was one which exhibited a broad molecular weight distribution (M_w/M_n=1.2) rather than samples prepared with this initiator which were narrow (M_w/M_n<1.1).

Hadjichristidis and coworkers [228] have investigated the use of initiator **90** for the polymerization of isoprene using a variety of ratios of [*sec*-BuOLi]/[CLi]. When this ratio was increased from 1 to 14, the polydispersity decreased from 1.20 to 1.03. In all cases the molecular weight distribution was monomodal. However, when this ratio was 8, a pale red color remained throughout the reaction, implying that unreacted initiator remained. When the ratio was 14, the red color disappeared and the viscosity was reduced as expected for the formation of cross-associated aggregates of the poly(butadienyl)lithium chain ends with the lithium *sec*-butoxide. These results confirm the earlier conclusion that the bimodal molecular weight distributions do not result from inexact stoichiometry [226], but are a consequence of chain end association effects [88].

8.2
Trifunctional Organolithium Initiators

By analogy with the structure of the *meta*-substituted double diphenylethylene, **73**, which forms a useful dilithium initiator upon addition of 2 moles of *sec*-butyllithium, the trifunctional diphenylethylene, **94**, has been investigated as a precursor for a hydrocarbon-soluble, trilithium initiator, **95**, as shown in Eq. (49) [239]:

$$3 \text{ } sec\text{-BuLi} + \mathbf{94} \longrightarrow \mathbf{95} \tag{49}$$

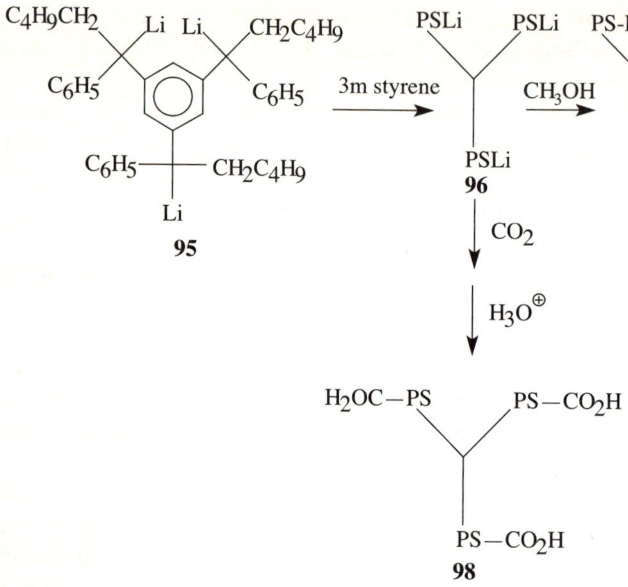

Scheme 32.

Like 73, 94 has 1-phenylethenyl units located *meta* to each other on a benzene ring which limits delocalization of negative charge among these substituents. The initiator precursor 94 undergoes quantitative addition of 3 equivalents of *sec*-butyllithium in benzene solution to form the trilithium initiator 95 [239]. The trilithium initiator was soluble in benzene solution at concentrations of 10^{-2} mol/l. However, analogous to the behavior of the dilithium initiator, 90 [88], when styrene was reacted with 95, incomplete crossover to monomer occurred after all of the styrene had polymerized [M_n(calc)=30,000]; residual absorption corresponding to the residual 1,1-diphenyl-alkyllithium initiator sites was observed at 450 nm. Since previous studies using the dilithium initiator, 90, suggested that inefficient initiation was the result of strong intermolecular association effects of the initiator with itself, the effect of addition of a Lewis base (THF) was investigated. Lewis bases such as THF promote dissociation of organolithium aggregates to form either unassociated species or aggregates with lower average degrees of association [3, 8, 19, 36, 113]. When the polymerization of styrene was initiated with 95 in the presence of THF ([THF]/[Li]=20), the UV absorption corresponding to the diphenylalkyllithium initiator sites completely disappeared. The resulting three-armed, star-branched polystyrene (97, see Scheme 32) exhibited a narrow (M_w/M_n=1.03) monomodal molecular weight distribution as determined by SEC and good agreement between calculated and observed number average molecular weight [239]. The resulting star-branched polystyrene also exhibited a g' value ([η]$_b$/[η]$_l$) of 0.81 which is good agreement with literature values for well-defined, three-armed star polymers [240]. Previous results have shown that g' is sensitive to arm heterogeneity [197, 207]. When the arms were not symmetrical, higher values of g' were observed.

The ω,ω',ω''-trilithiumpolystyrene (**96**) was also functionalized with carbon dioxide to form the corresponding ω,ω',ω''-tricarboxypolystyrene (**98**) as shown in Scheme 32. Although significant amounts of dimer (11–12%) were observed when carbonation was effected in the presence of THF [241] or after end-capping the styryllithium chain ends with 1,1-diphenylethylene [141], procedures that were previously shown to be effective for quantitative carboxylation of poly(styryl)lithium, less than 2% dimer formation and formation of a tricarboxylated, three-arm, star-branched polystyrene with a functionality of 2.9_5 were obtained upon carboxylation of a freeze-dried [141] sample of **96**. Previous studies indicate that dimer and trimer formation during carboxylation are enhanced by chain-end association [141, 241].

The polymerization of butadiene with the trifunctional organolithium initiator **95** in benzene resulted in complete crossover to butadiene when all of the monomer was consumed [242], in contrast to the analogous polymerization of styrene [239]. However, the SEC of the resulting polybutadiene exhibited a bimodal molecular weight distribution. As discussed for the dilithium initiator, **90**, it was proposed that the bimodality was due to aggregation effects and that added lithium alkoxide may promote uniform initiation reactions by forming cross-associated adducts (see **92**). When the polymerization of butadiene in benzene was initiated by **95** in the presence of lithium *sec*-butoxide ([LiOR]/[RLi]=0.5), the SEC chromatogram of the resulting polymer was monomodal. This polybutadiene exhibited a narrow molecular weight distribution [M_w/M_n(SEC)=1.03] and a g' value of 0.78. These results are consistent with formation of a regular, three-armed, star-branched polybutadiene using the trifunctional initiator, **95**.

Thus, multifunctional 1,1-diphenylethylenes, such as **73** and **94**, are precursors for useful, hydrocarbon-soluble, multifunctional organolithium initiators such as **90** and **95**, respectively. Their hydrocarbon solubility appears to be a consequence of a specific type of intermolecular association (e.g., **93**) which is favored over the more usual type of 3-dimensional association which leads to insolubility for most dilithium initiators [89, 220]. However, perhaps because of their unique association, these initiators require the addition of either a Lewis base, such as tetrahydrofuran, or a lithium alkoxide salt to initiate rapidly (relative to propagation) and quantitatively.

9
Multifunctional Living Linking Agents Based on 1,1-Diphenylethylene

Polymeric organolithium compounds react simply and quantitatively with 1,1-diphenylethylenes [3, 109]. These reactions have provided a new methodology for the synthesis of star-branched polymers, internally-functionalized polymer chains and stars, as well as heteroarm, star-branched polymers via living linking reactions as shown in Scheme 33 [3, 202, 203, 207].

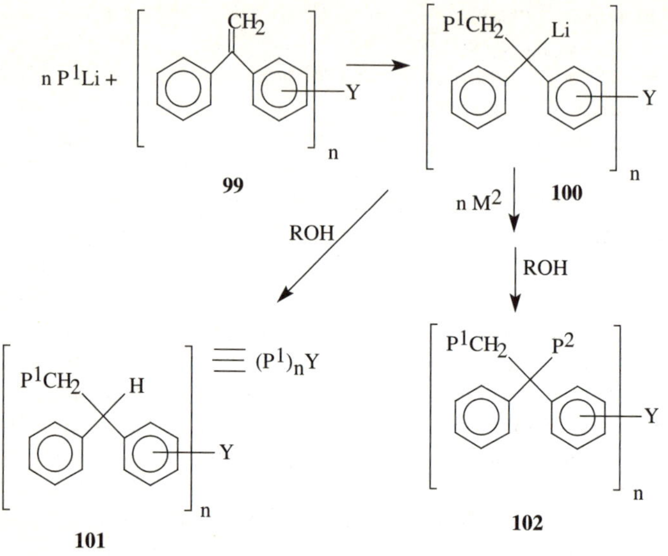

Scheme 33.

An anionic living linking agent (**99**) is a species which can react with polymers which have carbanionic chain-ends to generate a linked polymer product (**100**) which retains the active center stoichiometry as shown in Scheme 33 [3, 202]. If the linked product (**100**) is terminated with alcohol, the product (**101**) is either a coupled product (n=2) or a star-branched polymer (n>2). If the linked product (**100**) is reacted with an electrophilic functionalizing agent (E), the product (**103**) is either a coupled product (n=2) with two functional groups at the center of the chain or a star-branched polymer (n>2) with n functional groups at the core of the star (Eq. (50)) [206]:

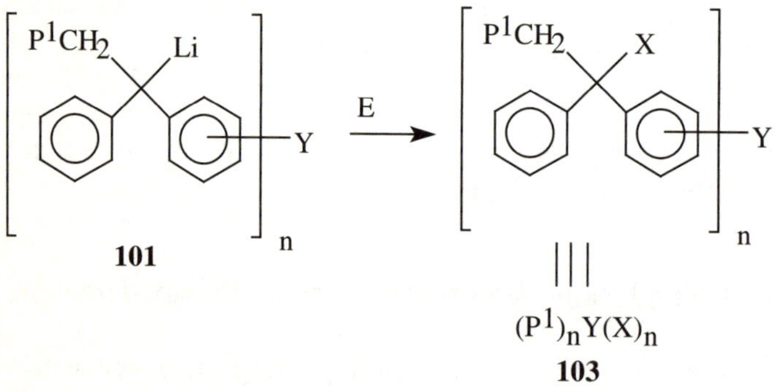

(50)

In addition, 1,1-diphenylalkyllithium sites in the living linked polymer product (**100**) can initiate polymerization of a second monomer (M^2) to generate a heteroarm (or mikto-arm), star-branched polymer, **102** [243, 244].

Several criteria must be satisfied for a species to be useful as a living linking agent [202]:
1. The living linking agent must react quantitatively with living carbanionic chain ends without oligomerization.
2. The coupled product must retain the active centers stoichiometrically.
3. The living coupled product must be capable of quantitatively reacting with electrophiles.
4. The living coupled product must be capable of reinitiating polymer chain growth rapidly (relative to propagation) and stoichiometrically.

Generally, living linking agents have two or more reactive vinyl substituents which can undergo addition reactions with polymeric carbanions, but which do not polymerize anionically due to steric hindrance. Divinylbenzenes and *m*-diisopropenylbenzene could be regarded as potential living linking agents; however, the homopolymerization and oligomerization, respectively, of these divinyl compounds limit their effectiveness for the synthesis of well-defined star-branched polymers [3, 207]. The most effective types of living linking agents are 1,1-diphenylethylene-based compounds as illustrated in Scheme 33. In the following sections, the scope and limitations of the use of 1,1-diphenylethylene-based living linking agents will be described.

9.1
1,3-Bis(1-phenylethenyl)benzene

With respect to the criteria enunciated for a useful living linking agent [202], 1,3-bis(1-phenylethenyl)benzene (MDDPE, **73**) is very effective. In terms of criterion 1 above, MDDPE reacts quantitatively with 2 equivalents of poly(-styryl)lithium in hydrocarbon solution at room temperature as shown in Eq. (51):

$$2\ PSLi + \mathbf{73} \longrightarrow \mathbf{104} \quad (51)$$

The coupling efficiency was examined by a combination of ultraviolet spectroscopy and size exclusion chromatography (SEC) [109, 202, 207]. These coupling reactions were monitored using UV-visible spectroscopy by observing the increase in absorbance of the corresponding diphenylalkyllithium species (**104**) at 438 nm. SEC was also useful to follow the course of the coupling reaction as shown in Fig. 3.

The coupling reaction of PSLi of various molecular weights with MDDPE in benzene is a very efficient reaction when the stoichiometry of the reaction is carefully controlled by determining the exact chain-end concentration [109,

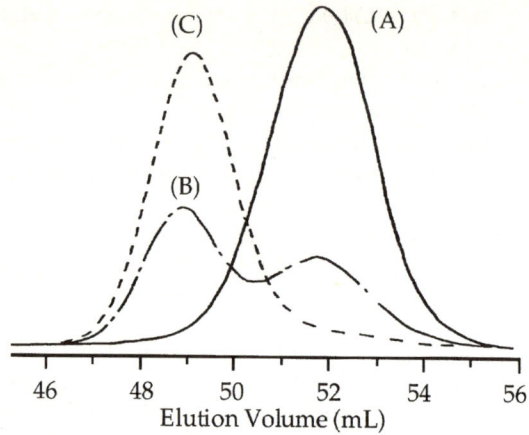

Fig. 3. SEC chromatograms of the coupling reaction of MDDPE with poly(styryl)lithium (M_n=1000 g/mol) in cyclohexane; base polymer (*A*); 55% coupling (*B*); and 100% coupling (*C*) (from [207]; reprinted with permission of Marcel Dekker)

Table 9. Molecular weight characterization data and coupling efficiencies in the reaction of stoichiometric amounts of PSLi with MDDPE in benzene as determined by SEC [109, 203, 207]

PS		(PS)$_2$		
M_n	M_w/M_n	M_n (coupled)	M_w/M_n	Coupling efficiency (%)
1,800	1.06	3,700	1.05	99
3,000	1.04	6,300	1.04	99
4,600	1.04	9,650	1.04	97
9,800	1.02	23,500	1.02	96
14,600	1.03	30,250	1.03	96

202, 207]. Coupling reactions of MDDPE with poly(styryl)lithium produced the coupled product in excellent yields (>96%) as shown in Table 9; the efficiency decreased slightly with increasing molecular weight.

The high reactivity of the second diphenylethylene unit in MDDPE with respect to formation of the coupled product (**104**) is further exemplified by the observation that, using cyclohexane as solvent at room temperature, the addition reaction of poly(styryl)lithium (M_n=3600) with a 3 molar excess MDDPE ([MDDPE]/[PLSi]=4) proceeds to give the dimeric adduct in 87% yield [197]. It should be noted that these results and previously published data [202, 203] stand in sharp contrast to the results of Ikker and Möller [206] who reported an inability to obtain high coupling efficiencies for the reactions of poly(styryl)lithium with MDDPE in benzene using the cryptand [2.1.1].

One limitation of using MDDPE as a living linking agent is the relative unreactivity of poly(dienyl)lithium compounds in addition reactions with 1,1-diphenylethylene units [109, 125]. It is necessary to add a small amount of THF ([THF]/[PLi]=40) or other Lewis base such as triethylamine to accelerate the addition reaction [109, 208]. The Lewis base is added after the polymerization of the diene is completed, so that the high 1,4-diene microstructure is preserved [3]. The coupling efficiency of poly(butadienyl)lithium ($M_n = 1.9 \times 10^3$; $M_w/M_n=1.03$) with MDDPE in cyclohexane was >99% in the presence of added THF ([THF]/[PLi]=40) [109, 208]. The coupled product exhibited $M_n=4.0 \times 10^3$ and $M_w/M_n=1.05$ by SEC analysis. Unfortunately, the required addition of a Lewis base to promote this dienyllithium coupling reaction limits the possibility of further diene polymerization using the living linked product (see Scheme 33) because only relatively high 1,2-diene microstructure, and consequently higher T_g diene blocks, would result [3]. A further complication with respect to the coupling reaction of poly(butadienyl)lithium with MDDPE is that unpolymerized butadiene monomer can cause the formation of small amounts of high molecular weight components in the coupled product because the unreacted butadiene will copolymerize with 1,1-diphenylethylene units to generate branch functionalities higher than two.

A further limitation with respect to the coupling reaction is the requirement that the living carbanionic polymers utilized must be sufficiently reactive to undergo facile addition reactions to 1,1-diphenylethylene units. In practice, this means that the first arms are limited primarily to styrene- and diene-type monomers.

9.1.1
Preparation of Heteroarm, Star-Branched Polymers

One of the most useful applications of living linking chemistry with 1,3-bis(1-phenylethenyl)benzene (**73**) and other multiple 1,1-diphenylethylenes is the synthesis of heteroarm star-branched polymers (**102**) by reacting the living linked products (**100**) with additional monomer as shown in Scheme 33. Of critical importance for the synthesis of heteroarm star-branched polymers was the determination of reaction conditions which would accelerate the rate of the crossover reaction of the diphenylalkyllithium sites with monomer (initiation) (k_i) relative to the rate of propagation (k_p), as described by the relative rate constant ratio, $k_i/k_p=R$. The values of R for crossover to styrene monomer in benzene were determined to be 0.07 and 0.10 at 25 °C and 5 °C, respectively (see Table 4) [109]. The analogous value of R at 25 °C in cyclohexane was 0.12 (see Table 4) [109]. Using the molecular weight distribution as an approximate experimental criterion for R, it was found that the narrowest molecular distributions for the polymers obtained by crossover experiments of polymeric diphenylmethyllithiums with styrene were obtained at 5 °C (vs 25 °C or 45 °C) in cyclohexane (vs benzene) [109, 207]. 1,1-Diphenyl-alkyllithiums can initiate styrene monomer polymerization with a reasonably rapid

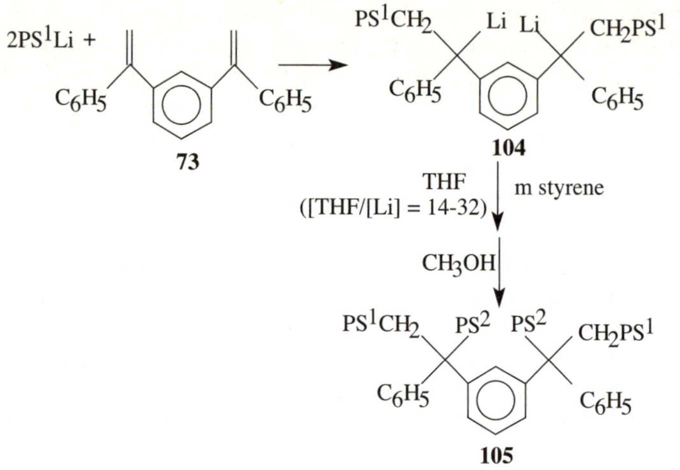

Scheme 34.

rate, even though it is slower than its homopolymerization rate (R<1). Stoichiometry is very important; when excess MDDPE was present in the crossover reaction, the star polymer product exhibited multimodal distribution by SEC analysis (with branching functionality greater than 4 arms) [245]. Young and coworkers [246] observed that high molecular weight contaminants were formed during the synthesis of styrene-isoprene, A_2B_2 heteroarm, star-branched copolymers. This was attributed to inexact stoichiometry during the linking reaction.

In order to investigate the best conditions for synthesis of heteroarm star-branched polymers using MDDPE, the preparation of four-armed, star-branched polystyrenes was examined in detail as illustrated in Scheme 34 [203, 207].

Even under stoichiometric coupling reaction conditions, the star-branched polymer (105) obtained from the crossover reaction with styrene to form growing polystyrene arms with $M_n=3.0\times10^3$ in each arm in cyclohexane solution showed bimodal molecular weight distribution [207]; the degree of bimodality reflects the relative number of molecules growing by propagation with two reactive sites versus one reactive site. Whenever bimodality was observed, the UV-visible spectrum showed the existence of residual absorption for unreacted diphenylalkyllithium sites at 438 nm. This situation is analogous to the bimodality observed with the dilithium initiator formed by reaction of sec-butyllithium with MDDPE (90) which was used to initiate the polymerization of styrene or butadiene monomer [88].

Addition of THF ([THF]/[Li]=14–32) to a coupled product (104) prior to addition of styrene produced narrow molecular weight distribution, star-branched polymers (105) even for polystyrene growing arm molecular weights as low as 3.0×10^3 as shown in Fig. 4. However, when the growing arm mole-

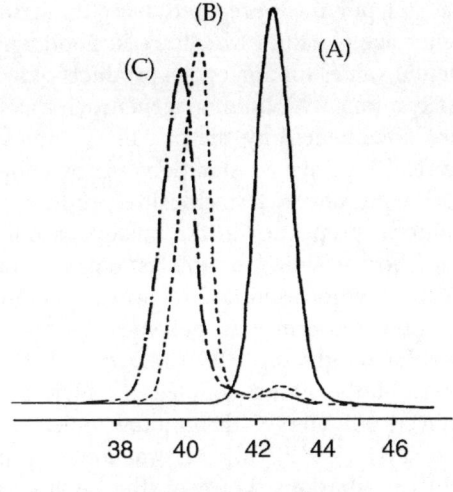

Fig. 4. SEC chromatograms of base polystyrene, M_n=12,000 g/mol (*A*); coupled product, (PS)$_2$-MDDPE (*B*); and (PS)$_2$-MDDPE-(PS)$_2$ star polymer with arm-out M_n (PS)=3000 g/mol (*C*) (from [207]; reprinted with permission of Marcel Dekker)

Table 10. Characterization of symmetrical, four-armed star polymers [203]

Polymer[a]	M_n Universal calibration[b]	M_n VPO[c]	$[\eta]_b$[d] (dl/g)	g'[e]
(3000)$_2$-(3000)$_2$	13×10^3	13×10^3	0.081	0.75
(5000)$_2$-(5000)$_2$	22.7×10^3	19.6×10^3	0.114	0.72
(10,000)$_2$-(10,000)$_2$	44.4×10^3	40.3×10^3	0.19	0.75
(10,000)$_4$Si[f]	44.4×10^3	38.3×10^3	0.19	0.75

[a] Numbers in parentheses indicate the calculated arm number average molecular weights.
[b] See [261].
[c] Vapor pressure osmometry.
[d] Intrinsic viscosity in THF.
[e] $g'=[\eta]_b/[\eta]_l$ where $[\eta]_l$ is calculated using the Mark-Houwink equation.
[f] Linking product of poly(styryl)lithium (M_n=10×10^3) with silicon tetrachloride.

cular weight was only 2×10^3, the presence of residual diphenylalkyllithium species was detected by UV-visible absorption at 438 nm [203]. Thus, the minimum arm length required for complete crossover to styrene monomer in the presence of THF is 3×10^3.

In order to investigate the reliability and usefulness of this star-branched polymer synthesis, a series of symmetrical, four-armed star polymers was prepared using the procedure outlined in Scheme 34. The results of the analyses of these polymers are shown in Table 10. It is perhaps most noteworthy that the symmetrical star polymers prepared using the living coupling procedure described herein have g' values which are in good agreement with both theory and experimental values for analogous products of symmetrical linking reactions of polymeric anions with linking agents such as silicon tetrachloride [203]. This is further documented by the fact that synthesis of an authentic sample of a symmetrical four-armed star polymer by coupling poly(styryl)-lithium ($M_n=10\times10^3$) with silicon tetrachloride produced a polymer which exhibits identical solution properties to the analogous polymer prepared by the living coupling method as shown by the last entry in Table 10. It is therefore concluded that the method outlined in Scheme 34 can be used for the controlled synthesis of heteroarm star polystyrenes with two set of arms which differ in molecular weight, e.g., $(PS^1)_2MDDPE(PS^2)_2$.

Crossover reactions of the living coupled product, **104**, with butadiene monomer also produced polymers with bimodal molecular weight distributions. The addition of THF ([THF]/[Li]=32) was found to produce monomodal molecular weight distributions; however, the polybutadiene microstructure was 42% 1,2 [88]. Diene microstructure is dramatically changed from high 1,4- to high 1,2-microstructure for alkyllithium-initiated diene polymerization in the presence of even small amounts of THF [247]. Thus, the addition of THF is not an acceptable procedure for preparation of elastomeric polydiene blocks. It was found that the addition of sufficient amounts of lithium sec-butoxide ([sec-BuOLi]/[PLi]=1) to the living coupled product (**104**) prior to addition of butadiene monomer produced monomodal, heteroarm, star-branched polymers even with relatively low molecular weights for the polybutadiene (PBD) arm which grow out from the coupled product; however the molecular weight distribution was somewhat broad. This effect of lithium alkoxides was first observed for the dilithium initiator formed by the addition of two moles of sec-butyllithium to MDDPE [88]. In the presence of less than one equivalent of lithium sec-butoxide ([LiOR]/[RLi]<1), bimodal molecular weight distributions were obtained. Polymer products with minimum PBD

Table 11. Molecular weight characteristics and microstructures of star-branched $(PS)_2$-MDDPE-$(PB)_2$ styrene-butadiene block copolymers prepared in cyclohexane in the presence of lithium sec-butoxide

Sample	$M_n\times10^{-3}$		M_w/M_n	Microstructure of PB block (mol %)	
	Calcd.[a]	^{1}H NMR[b]		1,4-	1,2-
S-1	2^2–2.5^2(9)	8.7	1.14	85	15
S-2	2^2–10^2(24)	24.5	1.05	90	10
S-3	10^2–11.5^2(43)	42.5	–	92	8

[a] Stoichiometric molecular weights for $(PS)^2(PBD)^2$.
[b] Calculated using the polystyrene molecular weight (SEC) and the relative integration areas for the aromatic protons relative to the vinyl protons from the diene units.

arm molecular weights ($M_n=10.0\times10^3$) exhibited a reasonably narrow molecular weight distribution without significant effect on polybutadiene microstructure as shown by the characterization data in Table 11.

There are fewer restrictions on the monomers that can be used in the crossover reaction than for the linking reaction. In essence, any monomer which undergoes living anionic polymerization can be added to the dilithium adduct (**104**) to grow arms and form heteroarm star-branched polymers with well-defined structures and low degrees of compositional heterogeneity. Since the growing arms are living, block copolymer arms can also be prepared.

9.1.1.1
Preparation of Heteroarm, Star-Branched Styrene-Butadiene Thermoplastic Elastomers

The effect of heteroarm branching on the properties of styrene-butadiene thermoplastic elastomers has been investigated by preparing heteroarm, star-branched analogs of thermoplastic elastomers using the synthetic scheme outlined in Scheme 35 [109, 248]. The first step in this sequence is the living coupling reaction [202] of two molecules of poly(styryl)lithium (PSLi, $M_n=15,000$) with MDDPE (**73**) to form the corresponding dianion, **104**. The coupling reaction can be monitored by UV-visible spectroscopy by observing the decrease in absorbance of PSLi at 334 nm and the increase in absorbance of the corresponding diphenylalkyllithium species at 438 nm [203]. Size exclu-

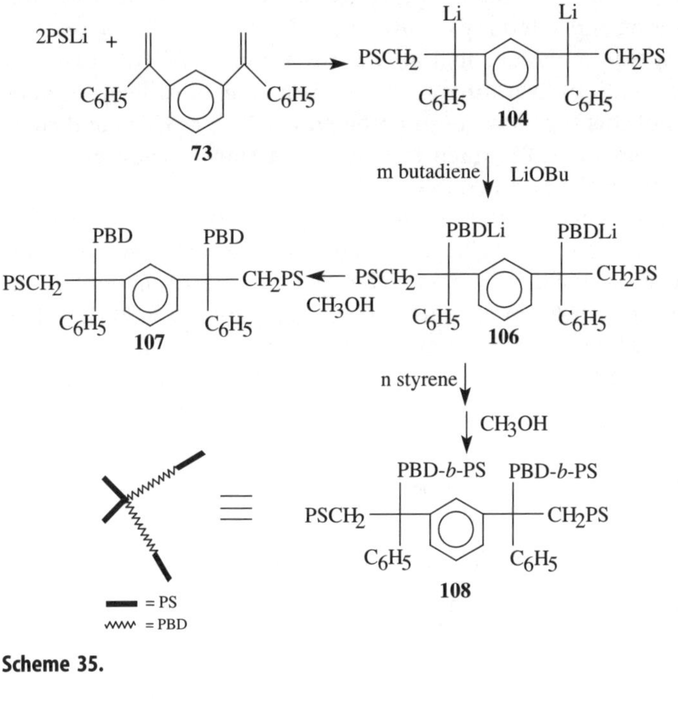

Scheme 35.

sion chromatography can also be used to evaluate the efficiency of the coupling reaction. These two probes generally indicate that the efficiency of this coupling reaction is essentially quantitative if stoichiometric quantities of reagents are utilized [203]. For example, the required amount of MDDPE is weighed using an analytical balance in a dry box.

In order to promote the efficient crossover reaction of the coupled product, **104**, with butadiene monomer, the addition of lithium alkoxide (lithium sec-butoxide; [LiOR]/[RLi]=1.0) was found to be useful analogous to the effect of lithium alkoxide with the dilithium initiator, **90** [88]. In the presence of lithium sec-butoxide, well-defined, monomodal, heteroarm, star-branched polymers (**107**) were obtained with high 1,4-microstructure of the polybutadiene blocks [203]. In the absence of the lithium alkoxide, bimodal molecular weight distribution polymers were obtained and residual UV absorption corresponding to the diphenylalkyllithium initiator groups at 438 nm was still observed after all of the monomer had been consumed.

In principle, the resulting living star-branched dianion, **106**, with poly(butadienyl)lithium arms can be reacted with styrene monomer to produce heteroarm, star-branched polymers, **108**. However, the rate of crossover of poly(butadienyl)lithium chain ends to styrene monomer is not rapid or competitive with respect to the ensuing styrene propagation in hydrocarbon media [3, 249–252]. As a consequence, it is difficult to obtain well-defined, narrow molecular weight distribution polystyrene blocks from these crossover reactions. In order to overcome this slow crossover reaction from poly(butadienyl)lithium chain ends to styrene monomer, it is necessary to add a small amount of a Lewis base, for example, tetrahydrofuran (THF) [252–254]. Thus, prior to the addition of styrene monomer to **106**, a small amount of THF (1 vol. %) was added to the solution. Two different heteroarm, star-branched, thermoplastic elastomers of this type have been prepared (ST70 and ST35) and compared with a linear analog (SBS); their structures are shown below and their characterization data are shown in Table 12 [109, 248].

No significant differences in the tensile properties of these three thermoplastic elastomers were observed. The ST35 polymer with higher styrene content exhibited an initial yielding behavior, a higher tensile strength at break (32 MPa), and longer elongation at break (700%) compared to ST70 (20 MPa, >800%) and SBS (20 MPa, >900%). Significant effects of branching were observed for the dynamic melt viscosities of these polymers measured at

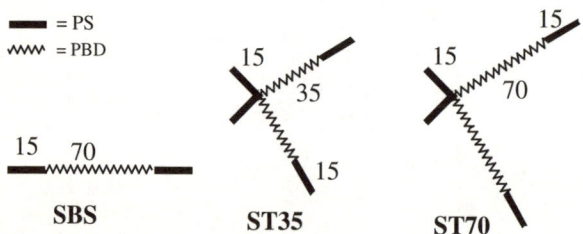

Table 12. Molecular characteristics of linear and star-branched thermoplastic elastomers

Sample	Calc $M_n \times 10^{-3}$ (total)	$M_n \times 10^{-3}$ (SEC)[a]	M_w/M_n (SEC-LALLS)	% styrene[b]	% vinyl[b]
SBS	15–70–15(100)	103	1.06	31	9
ST70	(15)2-(70–15)2(200)	199	1.08	31	14
ST35	(15)2-(35–15)2(130)	127	1.06	46	13

[a] Universal calibration method.
[b] ^{1}H NMR.

167.5 °C. Linear SBS and branched ST70 exhibited approximately the same melt viscosities even though ST70 has twice the molecular weight of SBS. ST35 exhibited a decade lower dynamic viscosity than either SBS or ST70. Thus, a useful effect of branching is to lower the melt viscosity of the star thermoplastic elastomers compared to the linear analog. In contrast, the intrinsic viscosities of these polymers vary in the order ST70 (150 ml/g)>SBS (106 ml/g)>ST35 (90 ml/g) [109]. Thus, ST70 exhibits a much higher intrinsic viscosity compared to SBS, reflecting more the higher molecular weight than the branching. However, ST35 exhibits a lower viscosity reflecting the lower hydrodynamic volume expected for a branched polymer with the same molecular weight as linear SBS.

9.1.2
Preparation of Functionalized Cyclic Polymers

The living linking chemistry of 1,3-bis(1-phenylethenyl)benzene (MDDPE, **73**) has been utilized to prepare cyclic polymers (**111**) and functionalized cyclic polymers (**110**) as shown in Scheme 36 [205, 255–257]. The commonly used

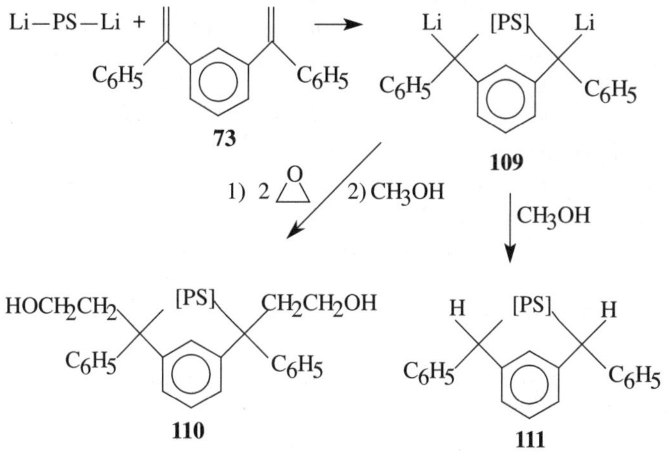

Scheme 36.

methods for synthesis of well-defined, cyclic polymers involve preparation of a monodisperse polymer initiated by a difunctional initiator and then reaction of the resulting polymeric α,ω-dianion in dilute solution with a difunctional, electrophilic coupling agent such as xylene dibromide or dimethyldichlorosilane [258, 259]. In contrast to these electrophilic reagents which terminate the active centers, the advantage of using a difunctional living coupling agent such as MDDPE (**73**) is that the reaction to form the macrocycle is not a termination reaction and the resulting product (**109**) retains the active center stoichiometry.

Lithium naphthalene was used to initiate the polymerization of styrene to generate α,ω-dilithiumpolystyrene in benzene as shown in Eq. (52):

$$2\ \text{LiNaph} + n\ \text{styrene} \longrightarrow \text{Li}\!-\!\!\text{[styrene]}_n\!\!-\!\text{Li} \quad (52)$$

Cyclization reactions were effected at room temperature in a glove box by dropwise addition of a solution of MDDPE (**73**) to a very dilute solution of α,ω-dilithiumpolystyrene as shown in Scheme 36 [255]. This scheme also illustrates the unique aspect of this macrocycle synthesis: the product of the cyclization reaction (**109**) is a living dianion which can be used for functionalization reactions to form functionalized macrocycles (**110**) [255, 256] or as a difunctional initiator to form macrocycles with branches [205, 257] (Eq. 53):

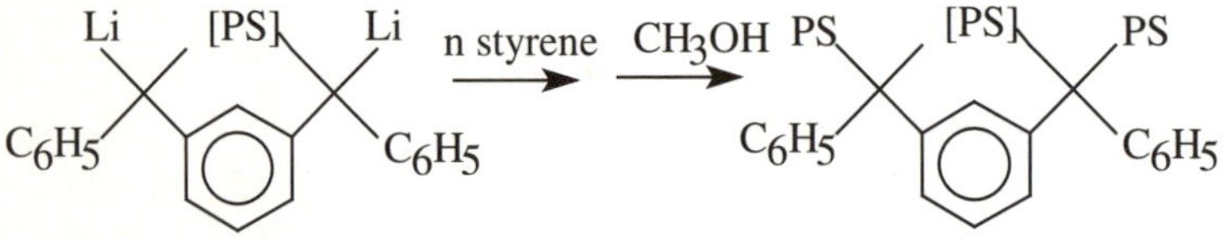

It was found that there was a major effect of solvent on this cyclization reaction involving the reaction of α,ω-dilithiumpolystyrene with MDDPE (**73**). In THF at –20 °C, the linear fraction was estimated to comprise 36–63% of the product mixture. However, when the cyclization reaction was effected in cyclohexane, only 2–3% linear contaminants were observed. This is consistent with previous studies of living linking reactions which showed that the reaction of two moles of poly(styryl)lithium with MDDPE produced only 19% dimer in THF, while the yield of dimer was essentially quantitative (>98%) in cyclohexane or benzene [197, 205].

The inefficiency of the cyclization reaction in tetrahydrofuran was indicated by the observation that, even after extensive fractionation, the ratios of the intrinsic viscosities of the cyclic to the linear polystyrenes (0.80; 0.82; 0.87) were quite high compared to the theoretical value of 0.67 [260] for polymers with M_n=4000. The amount of linear material contaminating the macrocyclic polymers was estimated by ultraviolet spectroscopy, assuming that linear contaminants (**113**) would have one unreacted diphenylethylene unit at the chain end (Scheme 37). The absorbance of the residual diphenylethylene chromo-

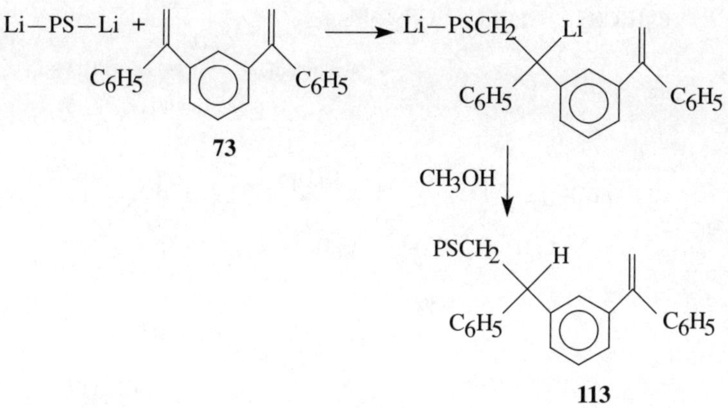

Scheme 37.

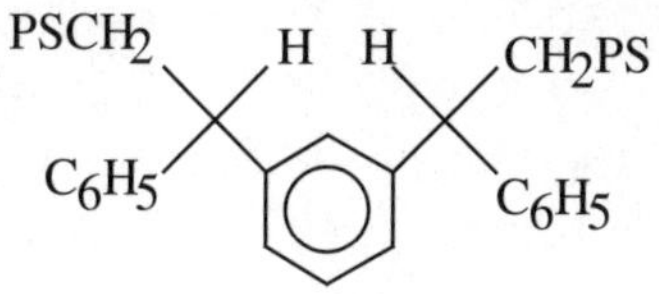

phore at 258 nm in the linear material has a molar absorptivity [$\varepsilon=1.18\times10^4$ (Mcm)$^{-1}$] that is 22 times larger than the absorbance of the diadduct of poly(styryl)lithium with MDDPE [**114**, $\varepsilon=540$ (Mcm)$^{-1}$] which is a model for the chromophore in the cyclic polymer.

UV analysis indicated that the macrocycles prepared in THF were contaminated with 30–50% of the linear polymer [205]. The presence of uncoupled 1,1-diphenylethylene units at linear chain ends was also detected by ^{13}C NMR analysis. The methylene carbon of the diphenylethylene unit exhibits a clearly observable resonance at δ 114 ppm which is present in the ^{13}C NMR spectra of the macrocycles prepared in THF, but this resonance is not detectable for samples of macrocycles prepared in cyclohexane.

The functionalized macrocyclic polystyrenes (**110**) prepared in cyclohexane exhibited ratios of intrinsic viscosities of the cyclic to the linear polymers of 0.74 and 0.76 for polymers with $M_n=6200$ and 4000, respectively, after fractionation [255]. The hydroxyl group functionalities were determined to be 2.0, 1.9, and 2.0 for macrocyclic polystyrenes with $M_n=4000$, 6200, and 12,200, respectively [205]. The macrocycle yields were approximately 40%.

Functionalized, macrocyclic polybutadienes with high 1,4-microstructure have been prepared using the dilithium initiator (**90**) prepared from MDDPE (**73**) as shown in Scheme 38 [205, 256]. A benzene solution of MDDPE (**73**)

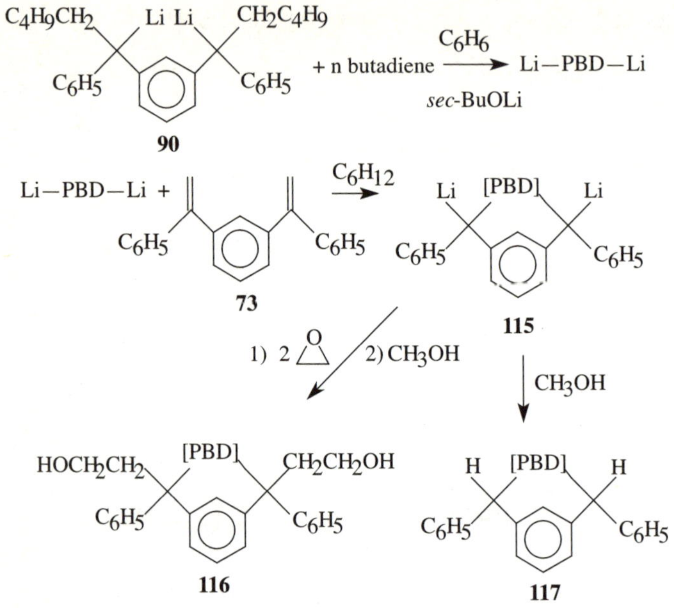

Scheme 38.

was added dropwise to a dilute solution of α,ω-dilithiumpolybutadiene in cyclohexane at 7 °C followed by functionalization with ethylene oxide to form the dihydroxy-functionalized macrocyclic polybutadiene (M_n=28,000) (**116**). The cyclization products were fractionated in a benzene/methanol mixture at room temperature. The intrinsic viscosity of the fractionated, functionalized, macrocyclic polybutadiene ([η]=46.0 ml/g) was much lower than that of an analogous (M_n=32,000) linear α,ω-dihydroxypolybutadiene ([η]=61.5 ml/g). The ratio of the intrinsic viscosity of the cyclic over the linear polymer was about 0.75, which is higher than the theoretical prediction of 0.66 [260]. One possible explanation for this difference is the presence of two rigid, double diphenylalkyl units introduced from the initiator and the living coupling agent. Both the cyclic and linear polybutadienes contained 30% *cis*-1,4-, 56% *trans*-1,4-, and 14% 1,2-microstructures. The glass transition temperatures for the functionalized linear and cyclic polymers were –88 °C and –86.7 °C, respectively, as measured by DSC.

10
Conclusions

1,1-Diphenylethylene (DPE) is a versatile adjuvant for anionic synthesis of well-defined polymers. DPE reacts relatively rapidly, quantitatively, and stoichiometrically with simple alkyllithiums and styryllithiums; however, the addition of dienyllithiums is too slow to be synthetically useful. It is necessary to add Lewis bases to promote the addition of dienyllithiums to DPE. The corresponding 1,1-diphenylalkyllithium from addition of simple or polymeric organolithiums with DPE is a more sterically hindered, less reactive initiator compared with alkyllithiums, styryllithiums, and dienyllithiums; however, it is useful for initiating polymerization of styrenes, dienes, and alkyl methacrylates.

Although DPE cannot homopolymerize, it will readily copolymerize (r_{DPE}=0) with styrenes; however, copolymerization with dienes requires the addition of a Lewis base promoter. Because DPE and its derivatives do not homopolymerize but are reactive toward addition of simple and polymeric organolithiums, substituted DPE derivatives are useful for introducing functional groups at the initiating chain end, within the polymer chain, at block junctions, and at the terminal chain end. DPE-functionalized macromonomers undergo simple, quantitative addition of polymeric organolithiums to provide methodologies for synthesis of ABC-type heteroarm, star-branched polymers. Hydrocarbon-soluble dilithium and trilithium initiators can be formed by stoichiometric addition of *sec*-butyllithium with multifunctional DPE derivatives such as bis(1,1-diphenylethylenes) and tris(1,1-diphenylethylenes), respectively. These multifunctional DPE derivatives are also useful as living linking agents for preparation of heteroarm, star-branched polymers.

Acknowledgements. RPQ would like to acknowledge Professor Michele Fontanille, Dr. Yves Gnanou, and Professor Alain Deffieux for their assistance and stimulating interactions during his stay as Visiting Professor at LÛniversité Bordeaux I in 1998, during which time most of the writing of this article was accomplished. RPQ would also like to acknowledge the careful reading and helpful suggestions of Keith Andes, Tae-Hee Cheong, Qing Ge, Kevin Jiang, Robert Mathers, Sergio Moctezuma, Myung-Ahn Ok, Joseph Pickel, Wenxin Yu, and Jin-Ping Zhou. The authors are grateful to the Dow Chemical Company and the National Science Foundation for financial support of much of our research described herein.

References

1. Bywater S (1985) In: Encyclopedia of polymer science and engineering, 2nd edn. Wiley-Interscience, New York, vol 2, p 1
2. Morton M (1983) Anionic polymerization: principles and practice. Academic Press, New York
3. Hsieh HL, Quirk RP (1996) Anionic polymerization: principles and practical applications. Marcel Dekker, New York
4. Szwarc M (1996) Ionic polymerization fundamentals. Hanser, New York
5. Fetters LJ, Thomas EL (1993) In: Material science & technology, vol 12. VCH, Weinheim, p 1–31
6. Szwarc M (1968) Carbanions, living polymers and electron transfer processes. Interscience, New York
7. Rempp P, Franta E, Herz JE (1988) Adv Polym Sci 86:145
8. Young RN, Quirk RP, Fetters LJ (1984) Adv Polym Sci 56:1

9. Szwarc M (1983) Adv Polym Sci 49:1
10. Bywater S (1974) Prog Polym Sci 4:27
11. van Beylen M, Bywater S, Smets G, Szwarc M, Worsfold DJ (1988) Adv Polym Sci 86:87
12. Morton M, Fetters LJ (1975) Rubber Chem Technol 48:359
13. Falk JC, Benedetto MA, Van Fleet J, Ciaglia L (1982) Macromolecular synthesis. Wiley, New York, vol 8, p 61
14. Richards RW, Thomason JL (1982) Polymer 23:1988
15. Stavely FW, Forster FC, Binder LL, Forman LE (1956) Ind Eng Chem 48:778
16. Szwarc M, Levy M, Milkovich R (1956) J Am Chem Soc 78:2656
17. Szwarc M (1956) Nature 178: 1168
18. Bywater S (1976) In: Bamford CH, Tipper CFH (eds) Comprehensive chemical kinetics: non-radical polymerization. Elsevier, New York, vol 15, p 1
19. Wakefield BJ (1974) The chemistry of organolithium compounds. Pergamon Press, New York
20. Ziegler K, Crossmann F, Kleiner H, Schafer O (1929) Liebigs Ann Chem 473:1
21. Köbrich G, Stöber I (1970) Chem Ber 103:2744
22. Evans AG, George DB (1962) J Chem Soc 141
23. Evans AG, George DB (1961) J Chem Soc 4653
24. Nakahama S, Hirao A, Ohira Y, Yamazaki N (1975) J Macromol Sci-Chem A9:563
25. Waack R, Doran MA (1971) J Organomet Chem 29:329
26. Waack R, Doran MA (1969) J Am Chem Soc 91:2456
27. Evans AG, Gore CR, Rees NH (1965) J Chem Soc 5110
28. Casling RAH, Evans AG, Rees NH (1966) J Chem Soc (B) 519
29. Waack R, Doran MA, Gatzke AL (1972) J Organomet Chem 46:1
30. Carpenter JG, Evans AG, Gore CR, Rees NH (1969) J Chem Soc (B) 908
31. Brown TL (1966) J Organomet Chem 5:191
32. Graham GD, Richtsmeier S, Dixon DA (1980) J Am Chem Soc 102:5759
33. Kaufmann E, Raghavachari K, Reed AE, von Rague Schleyer P (1988) Organometallics 7:1597
34. Seitz LM, Brown TL (1966) J Am Chem Soc 88:2174
35. Darensbourg MY, Kimura BY, Hartwell GD, Brown TL (1970) J Am Chem Soc 92: 1236
36. Fetters LJ, Young RN (1981) In: McGrath JE (ed) Anionic polymerization. kinetics, mechanisms, and synthesis. ACS Symposium Series No 166, American Chemical Society, Washington, D.C., p 95
37. Brown TL (1968) Accounts Chem Res 1:23
38 Lewis HL, Brown TL (1970) J Am Chem Soc 92:4664
39. Weiner J, Vogel C, West P (1962) Inorg Chem 1:654
40. Bywater S, Worsfold DJ (1967) J Organomet Chem 10:1
41. Worsfold DJ, Bywater S (1960) Can J Chem 38:1891
42. Brown TL, Gerteis RL, Bafus DA, Ladd JA (1964) J Am Chem Soc 86:2135
43. Bordwell FG (1988) Account Chem Res 21:456
44. Mathews WS, Bares JE, Bartmess JE, Bordwell FG, Cornforth FJ, Drucker GE, Margolin Z, McCallum RJ, McCollum GJ, Vanier NR (1975) J Am Chem Soc 97:7006
45. Bordwell FG, Fried HE (1981) J Org Chem 46:4327
46. Freyss D, Rempp P, Benoit H (1964) Polym Lett 2:217
47. Wiles DM, Bywater S (1964) Polym Lett 2:1175
48. Wiles DM, Bywater S (1965) Trans Faraday Soc 61:150
49. Fetters LJ, Morton M (1969) Macromolecules 2:453
50. Quirk RP, Zhu L (1990) Brit Polym J 23:47
51. Pauling L (1960) The nature of the chemical bond, 3rd edn. Cornell University Press, Ithaca, New York

52. Bock H, Ruppert K, Havlas Z, Bensch W, Honle W, von Schnering HG (1991) Angew Chem Int Ed Engl 30:1183
53. Patterman SP, Karle IL, Stucky GD (1970) J Am Chem Soc 92:1150
54. KcKeever LD (1972) In: Szwarc M (ed) Ions and ion-pairs in organic reactions. Wiley, London, p 263
55. McKeever LD, Waack R (1971) J Organomet Chem 28:145
56. Okamoto Y, Yuki H (1971) J Organomet Chem 32:1
57. Sandel VR, Freedman HH (1963) J Am Chem Soc 85:2328
58. Fraenkel G, Carter RE, McLachlan A, Richards JH (1960) J Am Chem Soc 82:5846
59. Zarges W, Marsch M, Harms K, Boche G (1989) Chem Ber 122:2303
60. Muller N, Pritchard DE (1959) J Chem Phys 31:768,1471
61. Wehrli FW, Marchand AP, Wehrli S (1983) Interpretation of carbon-13 NMR spectra, 2nd edn. Wiley, New York, p 66
62. van Dongen JPCM, van Dijkman HWD, de Bie MJA (1974) Recl Trav Chim Pays-Bays 93:29
63. Takahashi K, Kondo Y, Asami R (1974) Org Magnet Reson 6:580
64. Quirk RP, Monroy VM (1995) In: Kroschwitz JI (ed) Kirk-Othmer encyclopedia of chemical technology, 4th edn. Wiley, New York, vol 14, p 461
65. Müller AHE (1989) In: Eastmond GC, Ledwith A, Russo S, Sigwalt P (eds) Comprehensive polymer science, vol 3, chain polymerization 1. Pergamon, Oxford, UK, p 387
66. Müller AHE (1987) In: Hogen-Esch TE, Smid J (eds) Recent advances in anionic polymerization. Elsevier, New York, p 205
67. Piejko K-E, Höcker H (1982) Makromol Chem Rapid Commun 3:243
68. Hatada K, Kitayama T, Fumikawa K, Ohta K, Yuki H (1981) In: McGrath JE (ed) Anionic polymerization. kinetics, mechanisms, and synthesis. ACS Symposium Series 166, American Chemical Society, Washington, D.C., p 327
69. Anderson BC, Andrews GD, Arthur PJ, Jacobson HW, Melby LR, Playtis AJ, Sharkey WH (1981) Macromolecules 14:1599
70. Andrews GD, Melby LR (1984) In: Culbertson BM, Pittman CUJr, (eds) New monomers and polymers. Plenum Press, New York, p 357
71. Fayt R, Forte R, Jacobs C, Jerome R, Ouhadi T, Teyssie P, Varshney SK (1987) Macromolecules 20:1442
72. Varshney SK, Hautekeer JP, Fayt R, Jerome R, Teyssie P (1990) Macromolecules 23:2618
73. Kunkel D, Müller AHE, Janata M, Lochmann L (1992) Makromol Chem Macromol Symp 60:315
74. Janata M, Lochmann L, Müller AHE (1993) Makromol Chem 194:625
75. Wang J-S, Jerome R, Bayard P, Patin M, Teyssie P (1994) Macromolecules 27:4635
76. Zundel T, Zune C, Teyssie P, Jerome R (1998) Macromolecules 31:4089
77. Scherble J, Stark B, Stühn B, Kressler J, Budde H, Höring S, Schubert DW, Simon P, Stamm M (1999) Macromolecules 32:1859
78. Mecerreyes D, Dubois P, Jerome R, Hedrick JL (1999) Macromol Chem Phys 200:156
79. Quirk RP, Jang SH, Kim J (1996) Rubber Rev Rubber Chem Technol 69:444
80. Quirk RP, Jang SH, Han K, Yang H, Rix B, Lee Y (1998) In: Patil AO, Schulz DN, Novak BM (eds) Functional polymers. modern synthetic methods and novel structures. ACS Symposium Series No 704, American Chemical Society, Washington, DC, p 71
81. Bywater S (1989) In: Eastmond GC, Ledwith A, Russo S, Sigwalt P (eds) comprehensive polymer science, vol 3, chain polymerization 1. Pergamon, Oxford, UK, p 422
82. Quirk RP, Lee B (1992) Polym Int 27:359

83. Flory PJ (1953) Principles of polymer chemistry. Cornell University Press, Ithaca, New York, p 338
84. Flory PJ (1940) J Am Chem Soc 62:1561
85. Wang HC, Levin G, Szwarc M (1978) J Am Chem Soc 100:3969
86. Wang HC, Levin G, Szwarc M (1979) J Phys Chem 83:785
87. Szwarc M (1983) In: Bailey FEJr (ed) Initiation of polymerization. ACS Symposium Series 212, American Chemical Society, Washington, D.C., p 419
88. Quirk RP, Ma J-J (1991) Polym Int 24:197
89. Bandermann F, Speikamp H-D, Weigel L (1985) Makromol Chem 186:2017
90. Yuki H, Okamoto Y (1971) J Polym Sci Part A 9:1247
91. Alvarino JM, Bello A, Guzman GM (1972) Eur Polym J 8:53
92. Hsieh H (1965) J Polym Sci A 3:163
93. Margerison D, Newport JP (1963) Trans Faraday Soc 59:2058
94. Hsieh HL, Glaze WH (1970) Rubber Chem Technol 43:22
95. Busson R, van Beylen M (1977) Macromolecules 10:1320
96. Worsfold DJ, Bywater S (1960) Can J Chem 38:1891
97. Johnson AF, Worsfold DJ (1965) J Polym Sci Part A 3:449
98. Laita Z, Szwarc M (1969) Macromolecules 2:412
99. Al-Jarrah MM, Young RN, Fetters LJ (1979) Polym Prepr Am Chem Soc Div Polym Chem 20(1):739
100. Young RN, Fetters LJ, Huang JS, Krishnamoorti R (1994) Polym Int 33:217
101. Fetters LJ, Huang JS, Young RN (1996) J Polym Sci Part A Polym Chem 34:1517
102. Bywater S (1998) Macromolecules 31:6010
103. Bywater S (1995) Polym Int 38:325
104. Isaacs N (1995) Physical organic chemistry, 2nd edn. Longman Scientific and Technical, Essex, England, p 149
105. Burnett GM, Young RN (1966) Eur Polym J 2:329
106. Shima M, Bhattacharyya DN, Smid J, Szwarc M (1963) J Am Chem Soc 85:1306
107. Hammond GS (1955) J Am Chem Soc 77:334
108. Winstein S, Clippinger E, Fainberg AH, Robinson GC (1954) J Am Chem Soc 76:2597
109. Lee B (1991) PhD Dissertation, University of Akron, Akron, Ohio
110. Wang HC, Szwarc M (1980) Macromolecules 13:452
111. Zune C, Dubois P, Jerome R, Werkhoven T, Lugtenburg J (1999) Macromol Chem Phys 200:460
112. Bywater S, Worsfold DJ (1962) Can J Chem 40:1564
113. Morton M, Fetters LJ, Pett RA, Meier JF (1970) Macromolecules 3:327
114. Bhattacharyya DN, Lee CL, Smid J, Szwarc M (1965) J Phys Chem 69:612
115. Lochmann L, Lukas R, Lim D (1972) Coll Czech Chem Commun 37:569
116. Weidisch R, Michler GH, Fischer H, Arnold M, Hofmann S, Stamm M (1999) Polymer 40:1191
117. Zheng WY, Hammond PT (1998) Macromolecules 31:711
118. Yamada M, Itoh T, Nakagawa R, Hirao A, Nakahama S, Watanabe J (1999) Macromolecules 32:282
119. Yu JM, Dubois P, Teyssie P, Jerome R (1996) Macromolecules 29:6090
120. Johnson AF, Worsfold DJ (1965) J Polym Sci Part A 3:449
121. Yamagishi A, Szwarc M (1978) Macromolecules (1978) 11:504
122. Worsold DJ, Bywater S (1960) Can J Chem 38:412
123. Odian G (1991) Principles of polymerization, 3rd edn. Wiley-Interscience, New York, p 452
124. Mayo FR, Walling C (1950) Chem Rev 46:191
125. Yuki H (1972) Prog Polym Sci Jp 3:141
126. Yuki H, Hotta J, Okamoto Y, Murahashi S (1967) Bull Chem Soc Jp 40:2659

127. Trepka WJ (1970) Polym Lett 8:499
128. Ureta E, Smid J, Szwarc M (1966) J Polym Sci Part A-1 4:2219
129. Yuki H, Okamoto Y (1970) Polym J 1:13
130. Hatada K, Okamoto Y, Kitayama T, Sasaki S (1983) Polym Bull 9:228
131. Yuki H, Okamoto Y, Kuwae Y, Hatada K (1969) J Polym Sci Part A-1 7:1933
132. Hatada K, Kitayama T, Sasaki S, Okamoto Y, Masuda E, Kobayashi Y, Nakajima A, Aritome H, Namba S (1986) Ind Eng Chem Prod Res Dev 25:141
133. Yuki H, Okamoto Y (1970) Bull Chem Soc Jp 43:148
134. Yuki H, Okamoto Y (1969) Bull Chem Soc Jp 42:1644
135. Yuki H, Hatada K, Inoue T (1968) J Polym Sci Part A-1 6:3333
136. Yuki H, Okamoto Y, Sadamoto K (1969) Bull Chem Soc Jp 42:1754
137. Quirk RP (1992) In: Aggarwal SL, Russo S (eds) Comprehensive polymer science, 1st suppt. Pergamon Press, Oxford, p 83
138. Morton M, Fetters LJ (1967) Macromol Rev 2:71
139. Fetters LJ (1969) J Polym Sci Part C 26:1
140. Bywater S (1975) Prog Polym Sci 4:27
141. Quirk RP, Yin J (1992) J Polym Sci Part A Polym Chem 30:2349
142. Quirk RP, Kim J (1991) Macromolecules 24:4515
143. Durst T, Manoir JD (1969) Can J Chem 47:1230
144. Vanhorne P, Van den Bossche G, Fontaine F, Sobry R, Jerome R, Stamm, M (1994) Macromolecules 27:838
145. Quirk RP, Zhuo Q (1997) Macromolecules 30:1531
146. Quirk RP, Yin J, Guo S-H, Hu X-W, Summers GJ, Kim J, Zhu L-F, Ma J-J, Takizawa T, Lynch T (1991) Rubber Chem Technol 64:648
147. Quirk RP, Ren J (1993) In: Kahovec J (ed) Macromolecules 1992. VSP Publishers, Zeist, Netherlands, p 133
148. Heitz T, Höcker H (1988) Makromol Chem 189:777
149. Rempp PF, Franta E (1984) Adv Polym Sci 58:1
150. Kobayashi S, Uyama H (1994) In: Mishra MK (ed) Macromolecular design: concept and practice. Polymer Frontiers International, Hopewell Jct, New York, 1994, p 1
151. Yamashita Y (ed) (1993) Chemistry and industry of macromonomers. Huthig & Wepf, Heidelberg
152. Greene TW, Wuts PGM (1991) Protecting groups in organic synthesis, 2nd edn. Wiley-Interscience, New York
153. Schulz DN, Sanda JC, Willoughby BG (1981) In: McGrath JE (ed) Anionic polymerization. Kinetics, mechanisms, and synthesis. ACS Symposium Series No 166, American Chemical Society, Washington, DC, p 427
154. Patil AO, Schulz DN, Novak BM (eds) (1998) Functional polymers. Modern synthetic methods and novel structures. ACS Symposium Series No 704, American Chemical Society, Washington, DC
155. Hirao A, Nakahama S (1992) Prog Polym Sci 17:283
156. Nakahama S, Hirao A (1990) Prog Polym Sci 15:299
157. Hirao A, Nakahama S (1994) Trends Polym Sci 2:267
158. Quirk RP, Zhu L (1989) Makromol Chem 190:487
159. Hirao A, Kitamura K, Takenaka K, Nakahama S (1993) Macromolecules 26:4995
160. Quirk RP, Takizawa T, Lizarraga G, Zhu LF (1992) J Appl Polym Sci Polym Symp 50:23
161. Quirk RP, Yin J, Guo S-H, Hu X-W, Summers G, Kim J, Zhu L, Schock LE (1990) Makromol Chem Macromol Symp 32:47
162. Kim J, Kwak S, Kim KU, Kim KH, Cho JC, Jo WH, Lim D, Kim D (1998) Macromol Chem Phys 199:2185
163. Zhu L(1991) Ph D thesis, University of Akron
164. Quirk RP, Lynch T (1993) Macromolecules 26:1206

165. Lynch T (1996) Ph D thesis, University of Akron.
166. Hirao A, Ishino Y, Nakahama S (1988) Macromolecules 21:561
167. Summers GJ, Quirk RP (1996) Polym Internat 40:79
168. Hirao A, Nakahama S (1986) Polymer 27:309
169. Summers GJ, Quirk RP (1998) J Polym Sci Part A Polym Chem 36:1233
170. Quirk RP, Wang Y (1993) Polym Internat 31:51
171. Ishizone T, Hirao A, Nakahama S (1991) Macromolecules 24:625
172. Ishizone T, Sugiyama K, Hirao A, Nakahama S (1993) Macromolecules 26:3009
173. Ishizone T, Okazawa Y, Ohnuma K, Hirao A, Nakahama S (1997) Macromolecules 30:757
174. Quirk RP, Perry S, Mendicuti F, Mattice WL (1988) Macromolecules 21:2294
175. Chen L, Winnik MA, Al-Takrity ETB, Jenkins AD, Walton DRM (1987) Makromol Chem 188:2621
176. Liu G, Guillet JE, Al-Takrity ETB, Jenkins AD, Walton DRM (1990) Macromolecules 23:1393
177. Al-Takrity ETB (1995) Eur Polym J 31:383
178. Quirk RP, Schock LE (1991) Macromolecules 24:1237
179. Hruska Z, Vuillemin B, Riess G (1994) Polym Bull 32:163
180. Quirk RP, Kim J, Rodrigues K, Mattice WL (1991) Makromol Chem Macromol Symp 42/43:463
181. Wang Y, Kausch CM, Chun M, Quirk RP, Mattice WL (1995) Macromolecules 28:904
182. Quirk RP, Kim J, Kausch C, Chun M (1996) Polym Internat 39:3
183. Quirk RP, Ma J (1988) J Polym Sci Polym Chem Ed 26:2031
184. Calderara F, Hruska Z, Hurtrez G, Nugay T, Riess G, Winnik MA (1993) Makromol Chem 194:1411
185. Calderara F, Hruska Z, Hurtrez G, Lerch J-P, Nugay T, Riess G (1994) Macromolecules 27:1210
186. Hruska Z, Vuillemin B, Riess G, Katz A, Winnik MA (1992) Makromol Chem 193:1987
187. Ni S, Juhue D, Moselhy M, Wang Y, Winnik MA (1992) Macromolecules 25:496
188. Quirk RP, Zhu L (1992) Polym Internat 27:1
189. Quirk RP, Kuang J (1994) Macromol Symp 85:267
190. Hsieh HL, Wofford CF (1969) J Polym Sci A-1 7:449
191. Hsieh HL, Wofford CF (1969) J Polym Sci A-1 7:461
192. Quirk RP, Brittain WJ, Schulz GO (1996) In: Holden G, Legge NR, Quirk RP, Schroeder HE (eds) Thermoplastic elastomers, 2nd edn. Hanser, Cincinnati, p 395
193. Gnanou Y (1993) Ind J Technol 31:317
194. Percec V, Pugh C, Nuyken O, Pask SD (1989) In: Eastmond GC, Ledwith A, Russo S, Sigwalt P (eds) Comprehensive polymer science, vol 6, polymer reactions. Pergamon, Elmsford, New York, p 281
195. Quirk RP, Zhuo Q, Tsai Y, Yoo T, Wang Y (1995) In: Mishra MK, Nuyken O, Kobayashi S, Yagci Y, Sar B (eds) Macromolecular engineering: recent advances. Plenum Press, New York, p 197
196. Fujimoto T, Zhang H, Kazama T, Isono Y, Hasegawa H, Hashimoto T (1992) Polymer 33:2208
197. Quirk RP, Yoo T (1993) Polym Bull 31:29
198. Quirk RP, Yoo T (1993) Polym Prepr Am Chem Soc Div Polym Chem 34(2):578
199. Leitz E, Höcker H (1983) Makromol Chem 184:1893
200. Tung LH, Lo GYS (1994) Macromolecules 27:2219
201. Tung LH, Lo GYS (1994) Macromolecules 27:1680
202. Quirk RP, Ignatz-Hoover F (1987) In: Hogen-Esch TE, Smid J (eds) Recent advances in anionic polymerization. Elsevier, New York, p 393

203. Quirk RP, Lee B, Schock LE (1992) Makromol Chem, Macromol Symp 53:201
204. Broske AD, Huang TL, Allen RD, Hoover JM, McGrath JE (1987) In: Hogen-Esch TE, Smid J (eds) Recent advances in anionic polymerization. Elsevier, New York, p 363
205. Ma J-J (1991) PhD thesis, University of Akron
206. Ikker A, Möller M (1993) New Polymeric Mater 4:35
207. Quirk RP, Yoo T, Lee B (1994) J Macromol Sci-Pure Appl Chem A31:911
208. Quirk RP, Dixon H, Kim YJ, Yoo T (1996) Polym Prepr Am Chem Soc Div Polym Chem 37(2):402
209. Dixon H (1996) PhD thesis, University of Akron
210. Quirk RP, Kim YJ (1996) Polym Prepr Am Chem Soc Div Polym Chem 37(2):643
211. Kim YJ (1997) PhD thesis, University of Akron
212. Kim J, Cho JC, Kim KH, Kim KU, Jo WH, Quirk RP (1998) In: Patil AO, Schulz DN, Novak BM (eds) Functional polymers. Modern synthetic methods and novel structures. ACS Symposium Series No 704, American Chemical Society, Washington, DC, p 85
213. Huckstadt H, Abetz V, Stadler R (1996) Macromol Rapid Commun 17:599
214. Lambert O, Dumas P, Hurtrez G, Riess G (1997) Macromol Rapid Commun 18:343
215. Lambert O, Reutenauer S, Hurtrez G, Riess G, Dumas P (1998) Polym Bull 40:143
216. Yu JM, Dubois Ph, Teyssie P, Jerome R (1996) Macromolecules 29:6090
217. Long TE, Broske AD, Bradley DJ, McGrath JE (1989) J Polym Sci Part A Polym Chem 27:4001
218. Dong D, Hogen-Esch TE, Shaffer JS (1996) Macromol Chem Phys 197:3397
219. Foss RP, Jacobson HW, Sharkey WH (1977) Macromolecules 10:287
220. Tung LH, Lo GYS, Beyer DE (1978) Macromolecules 11:616
221. Guyot P, Favier JC, Uytterhoeven H, Fontanille M, Sigwalt P (1981) Polymer 22:1724
222. Lo GYS, Otterbacher EW, Gatzke AL, Tung LH (1994) Macromolecules 27:2233
223. Schulz G, Höcker H (1980) Angew Chem Int Ed Eng 19:219
224. Lo GYS, Otterbacher EW, Pews RG, Tung LH (1994) Macromolecules 27:2241
225. Tung LH, Lo GY (1986) In: Lal J, Mark JE (eds) Advances in elastomers and rubber elasticity. Plenum, New York, p 129
226. Bredeweg CJ, Gatzke AL, Lo GYS, Tung LH (1994) Macromolecules 27:2225
227. Bastelberger T, Höcker H (1984) Angew Makromol Chem 125:53
228. Iatrou H, Mays JW, Hadjichristidis N (1998) Macromolecules 31:6697
229. Roovers JEL, Bywater S (1968) Macromolecules 1:328
230. Roovers JEL, Bywater S (1966) Trans Faraday Soc 62:1876
231. McGarrity JF, Ogle CA (1985) J Am Chem Soc 107:1805
232. Brown TL, Ladd JA, Newman GN (1965) J Organomet Chem 3:1
233. Darensbourg MY, Kimura BY, Hartwell GE, Brown TIL (1970) J Am Chem Soc 92:1236
234. Narita T, Tsuruta T (1971) J Organomet Chem 30:289
235. Halaska V, Lochmann L (1973) Coll Czech Chem Commun 38:1780
236. Marsch M, Harms K, Lochmann L, Boche G (1990) Angew Chem Int Ed Eng 29:308
237. Ogle CA, Wang XL, Carlin CM, Strickler FH, Gordon B (1999) J Polym Sci Part A Polym Chem 37:1157
238. Yamagishi A, Szwarc M, Tung L, Lo GYS (1978) Macromolecules 11:607
239. Quirk RP, Tsai Y (1998) Macromolecules 31:8016
240. Grest GS, Fetters LJ, Huang JS, Richter D (1996) Star polymers: experiment, theory and simulation. Adv Chem Phys XCIV:67
241. Quirk RP, Chen W-C (1982) Makromol Chem 183:2071
242. Tsai Y (1995) Ph D thesis, University of Akron
243. Pitsikalis M, Pispas S, Mays JW, Hadjichristidis N (1998) Adv Polym Sci 135:1

244. Hadjichristidis N (1999) J Polym Sci Part A Polym Chem 37:857
245. Quirk RP, Schock LE, Lee B (1989) Polym Prep Am Chem Soc Div Polym Chem Div 30(1):113
246. Wright SJ, Young RN, Croucher TG (1994) Polym Internat 33:123
247. Antkowiak TA, Oberster AE, Halasa AF, Tate DP (1972) J Polym Sci Polym Chem Ed 10:1319
248. Quirk RP, Lee B (1991) Polym Prep Am Chem Soc Div Polym Chem Div 32(3):607
249. Morton M, Ells FR (1962) J Polym Sci 61:25
250. Hsieh HL, McKinney OF (1966) Polym Lett 4:843
251. Worsfold DJ (1967) J Polym Sci A-1 5:2783
252. Quirk RP, Morton M (1996) In: Holden G, Legge NR, Quirk RP, Schroeder HE (eds) Thermoplastic elastomers, 2nd edn. Hanser, Cincinnati, p 71
253. Morton M, McGrath JE, Juliana PC (1969) J Polym Sci Pt C 26:99
254. Fetters LJ, Morton M (1977) In: Macromolecular syntheses, collective vol 1. Wiley, New York, p 463
255. Quirk RP, Ma J-J (1988) Polym Prep Am Chem Soc Div Polym Chem Div 29(2):10
256. Quirk RP, Ma J-J (1992) Polym Prep Am Chem Soc Div Polym Chem Div 33(1):976
257. Ma J-J (1993) Polym Prep Am Chem Soc Div Polym Chem Div 34(2):626
258. Semlyen JA (ed) (1986) Cyclic polymers. Elsevier, New York
259. Rempp P, Strazielle C, Lutz P (1987) In: Encyclopedia of polymer science and engineering, vol 9, 2nd edn. Wiley-Interscience, New York, p 183
260. Bloomfield V, Zimm BH (1966) J Chem Phys 44:315
261. Grubisic Z, Rempp P, Benoit H (1967) J Polym Sci Polym Lett 5:753

Editor: Prof. H. Höcker
Received: February 2000

Device Applications of Polymer-Nanocomposites

D. Y. Godovsky

Energy and Semiconductor Department, University of Oldenburg, 26111 Oldenburg, Germany (*e-mail: dmitri.godovsky@uni-oldenburg.de*).

In recent years significant progress has been achieved in the synthesis of various types of polymer-nanocomposites and in the understanding of the basic principles which determine their optical, electronic and magnetic properties. As a result nanocomposite-based devices, such as light emitting diodes, photodiodes, photovoltaic solar cells and gas sensors, have been developed, often using chemically orientated synthetic methods such as soft lithography, lamination, spin-coating or solution casting.

Milestones on the way in the development of nanocomposite-based devices were the discovery of the possibility of filling conductive polymer matrices, such as poly(aniline), substituted poly(paraphenylenevinylenes) or poly(thiophenes), with semiconducting nanoparticles: CdS, CdSe, CuS, ZnS, Fe_3O_4 or fullerenes, and the opportunity to fill the polymer matrix with nanoparticles of both n- and p- conductivity types, thus providing access to peculiar morphologies, such as interpenetrating networks, p-n nanojunctions or "fractal" p-n interfaces, not achievable by traditional microelectronics technology.

The peculiarities in the conduction mechanism through a network of semiconductor nanoparticle chains provide the basis for the manufacture of highly sensitive gas and vapor sensors. These sensors combine the properties of the polymer matrix with those of the nanoparticles. It allows the fabrication of sensor devices selective to some definite components in mixtures of gases or vapors.

Magnetic phenomena, such as superparamagnetism, observed in polymer-nanocomposites containing Fe_3O_4 nanoparticles in some ranges of concentrations, particle sizes, shapes and temperatures, provide a way to determine the limits to magnetic media storage density, a problem which has been intensively investigated over the last five years.

Keywords. Nanocomposite photovoltaic solar cells, Polymer-nanocomposite light emitting diodes, Magnetic media storage capacity, Superparamagnetism, p-n Nanojunctions

1	Introduction	165
1.1	Polymer-Nanocomposites as Materials for Device Applications .	165
2	**Synthesis of Polymer-Nanocomposites for Device Applications** .	167
2.1	Preparation of Polymer-Composites Containing Both p- and n-Type Nanoparticles.	167
2.2	Preparation of Doped Nanoparticles	168
2.3	Gradient Composites Synthesis	169

2.4	Self-Assembly-Based Synthesis	170
2.5	Synthesis of Core-Shell Nanoparticles	170
2.6	Modification of the Surface of Nanoparticles	171
2.7	Gas-Phase Deposition Methods for the Preparation of Polymer-Nanocomposites	172
3	**Basic Physical Processes Determining the Operation of Polymer-Nanocomposite-Based Devices**	**173**
3.1	Quantum Confinement Effect in Polymer-Nanocomposites	173
3.2	Dipole–Dipole Interactions Between the Particles in Polymer-Nanocomposites	175
3.3	Percolation Behavior of Conductivity in Polymer-Nanocomposites	175
3.4	"Diffused" and "Fractal" p-n Junctions in Composites Filled with Both p- and n-Type Nanoparticles	177
4	**Electroluminescence in Polymer-Nanocomposites and Light Emitting Diodes**	**179**
4.1	Light Emitting Diodes Based on Polymer-Nanocomposites	182
5	**Photoconductivity and Photovoltaic Effect in Polymer-Nanocomposites**	**184**
5.1	Photodiodes and Solar Cells Based on Polymer-Nanocomposites	187
6	**Gas and Vapor Polymer-Nanocomposite Sensors**	**195**
7	**Polymer-Nanocomposites as Magnetic Storage Materials: Maximum Storage Density Problem**	**199**
8	**Conclusions**	**202**
9	**Future Perspectives of Nanocomposite-Based Devices**	**203**
References		**204**

List of Abbreviations

VLSI	very large scale integration circuits
LED	light emitting diode
Q-switch	quantum switch
PANI	polyaniline
ANGH	aniline hydrochloride
p-PPV	pristine polyphenylenevinylene
XRD	X-ray diffraction

PX   poly(p-xylylene)
MEH-PPV   methylethylhexoxypolyphenylenevinylene
L_D Debye length
EQE   external quantum efficiency
PT   poly(thiophene)
PMMA   poly(methylmethacrylate)
V_{oc} open circuit voltage
I_{sc} short circuit current
FF   fill factor
U_m voltage giving the maximal power output
I_m current giving the maximal power output
MIS   metal-insulator-semiconductor
PVA   poly(vinyl alcohol)
ITO   indium tin oxide
I-V   current-voltage
ChemFET   chemical sensor field effect transistor
AFM   atomic force microscopy
MFM   magnetic force microscopy

1
Introduction

1.1
Polymer-Nanocomposites as Materials for Device Applications

In recent years a second wave of interest in molecular electronics and organic compound based devices has taken place. The first wave in the early 1980s furnished the general understanding that the operating principles of molecular electronic devices are much more complicated than silicon-based ones, and that knowledge at that time was not sufficient to manufacture reliable long-term operable devices competitive with ones made by traditional microelectronics technology.

The 1990s brought into molecular electronics ideas of self-assemblies and the supramolecular engineering approach, which have changed the ideology of organic devices manufacture, opening up avenues to the production of a number of cheaper optoelectronic, photonic and electronic devices, thus attracting greater attention from photonics developers and IC producers.

The recent pioneering research by Cambridge Display Technologies Co., concentrated on the improvement of the long-term stability of organic light emitting diodes, has also shown [1] that significant improvements, using special sealing compounds and technological advances, can lead to reasonable operation times.

At the same time a change has taken place in the mentality of the developers of microelectronic devices who have recognized that, due to the rapid progress in the development of the hardware, the average life span of an electronic device such as a computer is now ca. three years. The other fact which has become more and more apparent is that the microminiaturization of transistor TTL or DTL based on very large scale integration circuits (VLSI) will soon reach its natural limit, and new ideology is needed for the next generation of computers and IT devices other than just scaling concepts, even though they have been applied successfully for almost 40 years.

Polymer electronics, including the use of polymer-nanocomposite-based devices, provide a number of alternative approaches, such as the use of adaptive circuits or the neural network-based processor architectures. Combined with a better understanding of the conductivity mechanism in conjugated polymers such as poly(acetylene) [2, 3] and poly(thiophenes) [4], these factors have initiated this second wave of interest in the low molecular weight organic and polymer-based optoelectronic, electronic and photonic devices.

Thus at present a wide range of polymer-based devices, such as light emitting diodes, photodiodes, solar cells, gas sensors, field effect transistors, exists which have been developed and intensively studied in research groups and R&D centers all over the world, some of which are even produced commercially in pilot scale series [5].

What are the advantages of the polymer-based devices that have attracted so much interest from the electronics manufacturers? The following lists some of them:

1 Low cost of production, usually by chemical synthetic methods rather than by microelectronics-based "clean" technologies;
2 Low cost of materials used for the device fabrication;
3 Possibility of facile and non-expensive manufacture of the large area devices such as LEDs or photovoltaic cells;
4 Availability of totally new material morphologies and device geometries, unattainable by traditional microelectronics technology methods;
5 Opportunity to make devices based on totally new principles such as bistable memory switches, neural networks, adaptive circuits, cellular automata, etc.;
6 Possibility to realize nanoscale size for the device structural elements without significant efforts especially using self-assembly-based techniques;
7 Enormous possibilities to vary the composition and hence properties of the organic and polymer materials used as device components.

All the advantages outlined make organic- and, in particular, polymer-based devices very attractive in our opinion at their present stage of development in microelectronics and photonics technology.

When speaking of nanocomposite-based devices, one should mention a significant feature which makes them different to fully organic-based ones: namely, the improved long-term stability. The stability of most polymer-nanocomposite devices exceeds that of fully organic-based devices dramatically, making them especially attractive for application purposes. It is particularly true for the nanocomposites made of inorganic nanoparticles in non-conductive polymer matrix based devices. This is due to the fact that the polymer matrix prevents to some extent interdiffusion of inorganic substituents, the latter being a weak point of some heterojunction-based inorganic semiconductor structures which often limits their operational time.

Another advantage of polymer-nanocomposites is the possibility of obtaining p-n nanojunctions, which are impossible in polymer-polymer systems, and, in particular, junctions between highly doped semiconductor particles, allowing the "fractalization" of the space charge layer, interpenetrating networks of nanoparticles and other valuable properties for device operation and peculiar material morphologies. Another consequence of the use of highly doped nanoparticles is the presence of high electric fields within a layer, which sometimes exceed 10^8 V/cm.

Besides the advantages outlined above, the opportunity to tailor the band gap simply by changing the size of the nanoparticles, representing such physical objects as quantum dots, provides an opportunity for the easy manipulation of LED photodiode characteristics. The light emission wavelength, for example, can be operated over a much broader range than by changing the stoichiometry of the binary or ternary chalcogenides traditionally used as inorganic LED materials.

Due to the fact that the size of nanoparticles can be made less than half that of the visible light wavelength they do not scatter light by Raleigh mechanism, thus the composite media can be almost as non-scattering as a single crystal. However, in contrast to single crystals, optical properties such as the refractive index can be tailored simply by changing the concentration of nanoparticles in the polymer matrix.

The other distinctive feature of polymer-nanocomposites is the extremely high interface area between the nanoparticles and the polymer matrix. This determines many of the specific properties, which can be both useful (as in sensor applications) and damaging for device operation.

2
Synthesis of Polymer-Nanocomposites for Device Applications

2.1
Preparation of Polymer-Composites Containing Both p- and n-Type Nanoparticles

We have developed [6] a number of methods to synthesize nanocomposites containing nanoparticles of both conduction types (p and n) in the same polymer matrix. These methods are discussed below.

(a) In Situ One-Sided Reactions. In order to obtain composites based on nonconductive matrices filled with semiconductor particles, in situ reactions conducted in water-swelled poly(vinyl alcohol) (host matrix) blended with polyacrylic acid (ionogenic agent) were carried out. The matrix was cross-linked to make it insoluble in water. A concentration of filler particles as high as 55 vol % was obtained using multi-cycle treatment. In order to obtain composites with two different sulfide species with different conductivity types within the same matrix, the reaction was carried out asymmetrically, i.e. from one side of the swelled film only, thus providing a concentration profile of one type of sulfide particles through the film thickness. As the next step, the swelled film was treated from the opposite side with the reagent to produce the other sulfide species, thus forming layers containing particles of only one conductivity type (e.g. CdS and Cu_2S) close to the edges and an intermixed layer containing particles of both types in the middle of the film.

(b) p-n Particles Conversion Reaction. Another method which has been used is the chemical conversion of one sulfide species into another, carried out from one side of the solution-swelled composite film:

$$CuCl + CdS \rightarrow Cu_2S + CdCl_2 \tag{I}$$

This reaction allowed the formation of an intermixed layer in the middle of the film and layers containing particles of only one sulfide species (CdS or Cu_2S) at the edges of the composite film.

To obtain conductive thin-film composite films, in situ polymerization of polyaniline (PANI) was carried out by means of oxidation of aniline hydrochloride (ANGH – monomer).

The other synthetic method applied was casting or spin-casting of colloid solutions of sulfide nanoparticles of both conductivity types together with the dissolved conductive polymer (PANI, p-PPV). This method produced composites with morphology similar to interpenetrating particle networks.

2.2
Preparation of Doped Nanoparticles

For a number of applications (described below), particles with high intrinsic conductivity, or more accurately high carrier concentration, are a necessity. This can be achieved by synthesizing donor or acceptor impurity doped semiconductors, since doping can alter the carrier concentration within the broad limits. We have succeeded in doping CdS nanoparticles with indium according to reaction scheme (II):

$$Cd(NO_3)_2 + Na_2S \rightarrow CdS_{nano} + 2NaNO_3 \tag{II}$$

$$InI_3 + Na_2S \rightarrow In_2S_3 + NaI \tag{III}$$

The conductivity of the composites containing indium-doped CdS particles increased more than 3 orders of magnitude in comparison with undoped ones.

Another way to control the carrier concentration is to vary the stoichiometry of sulfides such as Cu_2S or CuS. Since the conductivity of Cu_2S and CuS is determined by the excess of copper ions above the stoichiometry, controlling the ratio of the reagents during the synthesis can change the nanocomposite conductivity within the broad limits.

2.3
Gradient Composites Synthesis

We have introduced [7] the concept of photovoltaic cells and photodiodes based on the gradient of the concentration of the filler particles within the matrix, as shown in Fig. 1. The existence of this gradient allows optimization of the topology of the conductive particles network, preventing the formation of dead ends and hence decreasing the chance of recombination of the charge carriers during their transfer to the electrodes.

The gradient device morphology allows a better match to the built-in electric field, allowing a larger area of the device to be the active layer in a number of optoelectronic and photonic applications, thus increasing the amount of electron-hole pairs which can be effectively separated. Synthetic methods used to prepare the gradient structures are listed below.

a) One-sided in situ reactions [7].
A one-sided in situ synthetic method was developed which allows a gradient of nanoparticles within a polymer matrix. The reaction is carried out in the swelled cross-linked host polymer matrix with reagents having access to one side of the film only. Since the swelled matrix hinders the diffusion of the reagents, decay of the concentration of nanoparticles throughout the film is obtained.

b) Spin casting of nanocomposite solution onto the top of the swelled polymer layer followed by partial diffusion.
Another way to obtain a gradient of particle concentration consists of spin casting or solution casting the nanoparticles in a colloidal solution on top of the spin cast polymer film. Diffusion of nanoparticles into the polymer matrix takes place with a penetration depth that is controlled by the temperature. To enhance the diffusion solvents can be used that swell the polymer layer, though not dissolving it. By varying the spin and solution casting conditions various diffusion profiles can be obtained.

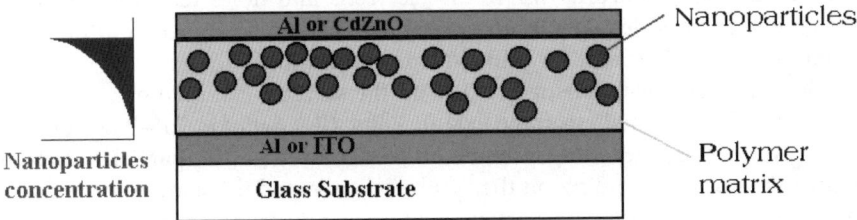

Fig. 1. Device geometry of a gradient-based structure

2.4
Self-Assembly-Based Synthesis

A method to self-assemble molecular electronic devices has been developed in our laboratory [8], consisting of stratification, i.e. spontaneous separation of polymer and nanoparticles when spin coating from a common bicomponent solution. The stratification leads to spontaneous formation of polymer layers with different concentrations of nanoparticles. The stratification is realized by mixing both dissolved polymer and the colloid solution, containing nanoparticles to be stratified with two different solvents, each solvent possessing different solubility towards each component and different boiling points.

In the process of casting or spin casting of the mixed solution, one of the solvents is evaporated faster leaving the solvent with the lower vapor pressure. The selective solubility therefore forces the precipitation of one of the components. This spontaneous separation of two components on the scale of 10–100 nm is usually followed by phase separation on the nanometer scale in the intermixed region and should allow complex morphologies due to non-equilibrium solid phase formation. This stratification produces diffused p-n junctions or interpenetrating donor-acceptor networks with a gradual connectivity change.

2.5
Synthesis of Core-Shell Nanoparticles

The synthesis of core-shell nanoparticles which possess a number of unique properties can be realized in several ways. Here we will describe one of them, probably the most illustrative, which was communicated by Fendler et al. [9]. The method comprises the injection of a gaseous mixture of H_2S and H_2Se into a solution of $Cd(ClO_4)_2$.

Depending on the ratio of H_2S to H_2Se, and the sequence in which the gases are injected, a number of structures can be obtained (Fig. 2). Some of these appear to be domain-like (b,c) while others are really core-shell ones (d,e). Since the confinement potential is highly dependent on whether the lower band gap material forms a core or a shell of the particle, a number of properties are completely different for a CdS/CdSe core-shell as opposed to CdSe/CdS. It especially concerns where the electrons and holes tend to rest, once the core-shell particles are placed in contact. The latter is crucial both for the manufacture of LEDs and for photovoltaic applications.

The oxide layer which usually exists on the surface of sulfide nanoparticles makes them "core-shell" to some extent. Thus XRD data for CdS-PVA composites give evidence for the existence of a CdO layer on the particles surface with a thickness depending on the synthetic route and the nature of the polymer matrix. Composite properties such as conductivity and a number of others can be affected to a large extent by the presence of this oxide surface

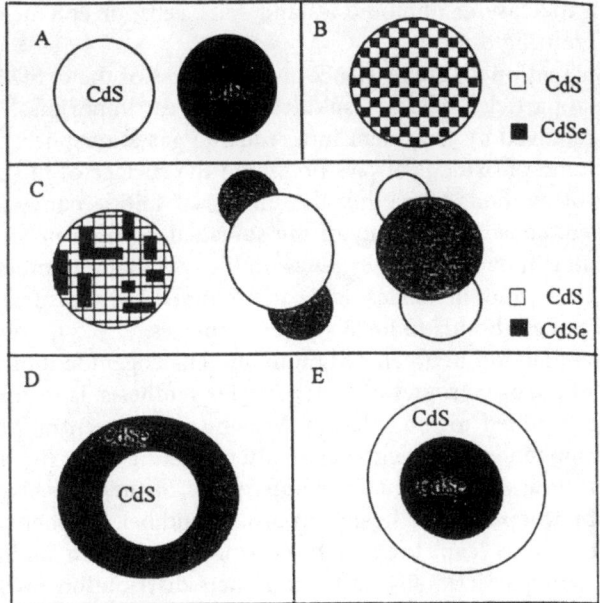

Fig. 2. Oversimplified diagram [9] of the possible structures of mixed CdS-CdSe nanoparticles. *A* Distinct and well-separated CdS and CdSe nanoparticles, *B* Well-mixed (solid) solutions of CdS-CdSe nanoparticles containing domains of CdS and CdSe, *C* Non-ideally mixed CdS-CdSe nanoparticles containing domains of CdS and CdSe, *D* Sandwich CdS-CdSe nanoparticles, *E* CdSe encapsulated CdS, CdS-core CdSe-shell type nanoparticles (the cherry and its stone), and *F* CdS encapsulated CdSe, CdSe-core CdS-shell type nanoparticles (the cherry and its stone)

layer. This is especially true for metal nanoparticles, and we observed the formation of an oxide layer for iron and silver nanoparticles which took place when the composites were subjected to ambient atmosphere. The time scale of oxide formation is several weeks and after a month a significant amount of nanoparticles is oxidized, leaving only a small inner core of metal. As a consequence, changes to the electrophysical, optical and magnetic properties of nanocomposites occur.

2.6
Modification of the Surface of Nanoparticles

The surface of nanoparticles is of great importance for a number of device applications, since a number of surface defects, such as dangling bonds or deep surface traps and dislocations, determine the details of the electron-hole recombination process, a process which plays a crucial role in optoelectronic devices operation. The trapping of electron-hole pairs must be either suppressed, as in the case of photodiodes and solar cells, or enhanced, as in the case of light emitting diodes.

As already mentioned, the existence and thickness of the oxide layer on the surface of nanoparticles made of sulfides is of great importance. This oxide layer can be removed by treatment in a reducing gas atmosphere, but usually a residual amount of oxide is always present at the surface of the particles.

A number of methods to modify the surface of sulfide nanoparticles have been developed, consisting mainly of the so-called capping procedure or organic ligand attachment. The latter relies on the fact that a number of organic molecules such as thiophenolates, etc. can be bonded to the surface of nanoparticles by covalent bonds to form stable complexes.

Actually the reaction between cadmium and chalcogenide ions in solution, which is one of the basic routes of nanoparticles synthesis, is terminated when thiophenolate is added to the solution. Varying the concentration of the reagents and thiophenolate, which acts as a terminating agent in the particles formation reaction, one can obtain nanoparticles of the desired size covered or "capped" by thiophenolate ligand groups. A number of other organic molecules can be used to "cap" the surface of nanoparticles in a similar way. The "capping" of nanoparticles also influences their distribution in the polymer matrix, another property vital for the manufacture of devices.

2.7
Gas-Phase Deposition Methods for the Preparation of Polymer-Nanocomposites

A number of methods have been developed recently based on the gas-phase deposition of monomers on a substrate followed by polymerization. These methods are particularly useful for device application, since they allow mask deposited circuits and can yield fairly thin composite films (down to 50 nm) on different substrate types allowing the polymer/filler ratio to be changed from 0 to 1, as well as giving access to peculiar composite morphologies by varying the synthetic conditions [10].

The first method to mention is the gas-phase deposition of organometallic monomers [11] such as p-cyclophanes with organogermanium or organotin substituents. The pyrolysis of the deposited compounds occurs without dissociation of the organometallic bonds and yields the corresponding p-xylylene monomers with organometallic substituents which, after deposition and polymerization, form poly(p-xylylenes) (PX) with organometallic units. Thermal treatment of the latter in an inert atmosphere causes breaking of the organometallic bonds and formation of PX composites with inclusion of Ge or Sn particles.

The other method is the co-condensation of metals and monomers onto the same substrate [12] followed by a polymerization step. The co-condensation of vinyl monomers with the vapors of metals such as Pd, Ag, Zn, Cd, Ga, In, Sn, Sb and Bi in a vacuum onto a cooled (77 K) substrate produces a condensate which is polymerized by heating or under the action of the usual radical initiators. It has been reported [13] that the co-condensation of gold and diacetylene allowed production of composites of gold particles in p-conjugated systems.

Polymer-metal composites can also be produced by means of polymerization in plasma (glow discharge) with the initiation of the reaction by high energy ions [14].

Cryochemical synthesis [15] is another recently developed method for producing thin films of metal and semiconductor filled polymer-nanocomposites with morphologies that are unattainable by other methods. The method consists of the light-induced polymerization of the co-condensate of the metal or semiconductor clusters with p-xylylene monomers which are highly reactive in the solid state. The application of light allows the degree of p-xylylene polymerization to be controlled quantitatively. Composites containing nanoparticles of metals, such as Mg, Pd, and semiconductors, such as PbS and CdS, with particle sizes and distribution depending on the polymerization conditions have been prepared using this method [16].

3
Basic Physical Processes Determining the Operation of Polymer-Nanocomposite-Based Devices

Polymer-nanocomposites with semiconductor nanoparticles possess a number of unique features which attracted much attention from physicists and chemists in the early 1980s [17–19]. Nanoparticles of semiconductors in a polymer matrix are extremely interesting to physicists since they exhibit behavior typical for one-dimensional electron gas systems with effects such as quantum confinement, discretization of the energy spectrum, unique non-linear optical properties, etc.

More detailed information can be found in the review articles mentioned above or in a review by the author in a previous volume of this series [19]. Here we will consider the quantum confinement effect briefly, since it plays a crucial role in nanocomposite-based device operation.

3.1
Quantum Confinement Effect in Polymer-Nanocomposites

Quantum confinement effects become "visible" when the size of the nanoparticle, and hence the size of the potential well which entraps or "confines" the electrons or holes within the particle, is comparable with the de Broigle wavelength of the electron or hole. In this case the electrons or holes in such a potential well always "feel" that they are confined. One of the main consequences of this fact is an increase in the conduction band electron or hole energy by a term known as the "particle-in-a-box" quantum energy E_i:

$$E_i = E_o + \frac{\pi^2 h^2}{2mR^2} \tag{1}$$

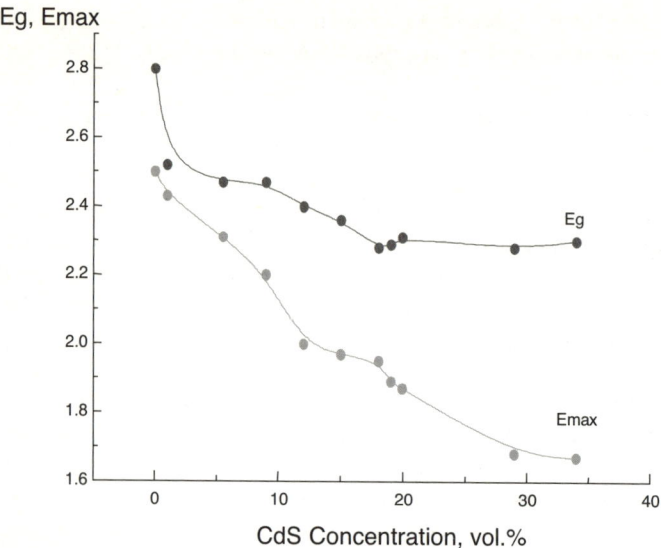

Fig. 3. Band gap of CdS-PVA nanocomposites determined from optical absorbance spectra (*upper curve*) and position of maximum of the photoluminescence peak (*lower curve*) depending on CdS concentration

where E_o is the initial energy of the particle, m is the effective mass of the electron or hole, and R is the radius of the nanoparticle. One consequence of this fact is an increase in the band gap if the particle size becomes smaller than the exciton radius. This phenomenon allows the band gap of a system to be tailored by changing the size of the particles, a property of great importance for many photonic and optoelectronic device applications.

We have observed [20] quantum confinement effects in optical absorbance spectra as well as on the photoluminescence peak position dependencies from filler concentration for CdS nanoparticles in a polyvinyl alcohol matrix (Fig. 3-a,b). The reason why these quantum confinement effects can be observed at filler concentration dependencies originates from the peculiarities of the in situ synthesis: the lower the concentration of nanoparticles produced by an in situ reaction, the smaller the average size of the particles.

Data from the P. Alivisatos group show another example of quantum confinement effects in the optical absorbance, photoluminescence and electroluminescence spectra of CdS and CdSe-MEH-PPV nanocomposites (Fig. 8). The mentioned phenomena can be utilized both for LED applications, as will be described below, or for the manufacture of photovoltaic cells. In the case of solar cells, filling the polymer matrix with nanoparticles which possess different band gaps makes them similar to multi-band gap solar cells, which were reported to show much higher values of energy conversion efficiencies than their single band gap counterparts. The effect becomes even more pronounced if the particles can be placed with the higher band gap closer to the illuminated side of the cell, thus allowing high energy photons to be absorbed in the upper layers, and the lower energy ones in the deeper laying layers of the cell.

3.2
Dipole–Dipole Interaction Between the Particles in Polymer-Nanocomposites

Another anomaly exists in the graph mentioned in the previous section depicting the band gap values on concentration dependencies as determined from optical absorbance measurements (Fig. 3a). In the area of high nanoparticles concentration the band gap values become smaller than the bulk values of the crystalline CdS.

One explanation for this anomaly, proposed by our group, assumes a strong interaction between dipoles induced in the neighboring nanoparticles analogous to van der Waal interactions in cases of molecule–molecule induced dipoles. Since for the range of concentrations under consideration the distances between nanoparticles are of the same order as the nanoparticle size (i.e. 1–5 nm) the effect can be quite strong and calculated local field values which are the result of such interactions exceed 10^7 V/cm. Such local fields can provide an effective narrowing of the band gap due to the Franc-Keldysh effect [20]. Calculated values of the band gap shift are in good agreement with those determined experimentally. The described model thus predicts the existence of strong local electric fields between nanoparticles, which is important when one considers the operation of devices such as photovoltaic cells or photodiodes based on polymer-nanocomposites.

Another effect discovered by our group in 1995 [20] is quenching of luminescence with the increase of CdS nanocrystal content in a poly(vinyl alcohol) (PVA) matrix, which is non-conductive and non-luminescent. This phenomenon will be described in more detail in Sect. 5.

3.3
Percolation Behavior of Conductivity in Polymer-Nanocomposites

Another fundamental phenomenon, which determines the behavior of conductivity in polymer-nanocomposites, is the percolation threshold, which occurs when the nanocrystals concentration reaches values high enough to provide conduction along the connected chains of nanoparticles. At the threshold values of filler concentration the conductivity of composite changes abruptly; the absolute value of this change can reach 10^7–10^8 Ohm/cm, which makes this effect similar to the metal-insulator transition in doped compensated semiconductors (Mott transition).

The author together with V. Sukharev has shown [21] that the percolation behavior of nanocomposites conductivity is different from the one typical for composites containing larger particles (Fig. 4). It has been demonstrated that the threshold filler concentration values are lower for nanocomposites than for composites with micron-size particles, and the slope of the curve in the

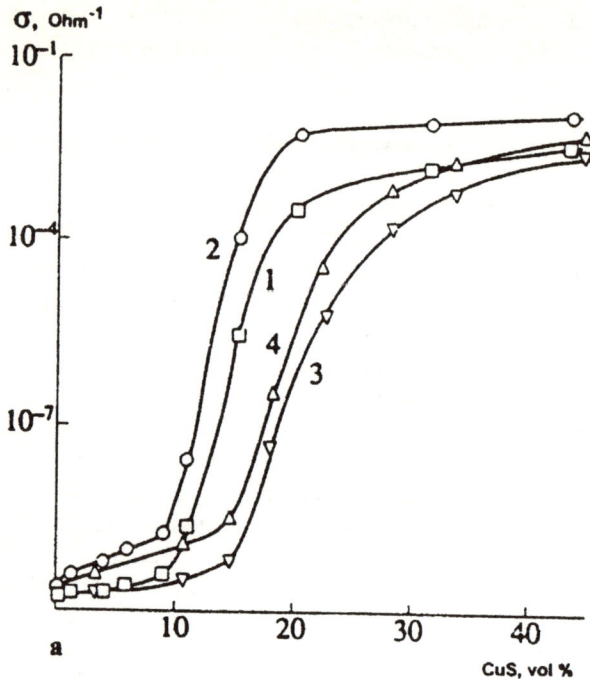

Fig. 4. The dependence of conductivity on concentration for nanosize (curves 1–5 nm and 2–12 nm) and micron size (curve 3–6 m, curve 4–10 m) composites (the average particle size is taken, a log-normal distribution of particles sizes was observed)

threshold area is steeper. The phenomenon originates from the presence of a filler concentration range where the conduction is dominantly of hopping origin. Hopping conduction has many peculiarities [22] which can be utilized for the manufacture of gas and vapor sensors (see Sect. 6).

The percolation-like behavior of polymer-nanocomposites conductivity is a crucial property for device manufacture since all the electrophysical and a number of other properties change completely once the percolation threshold concentration is reached. The question should always be asked whether to utilize the properties of non-contacting nanoparticles in a polymer matrix below the threshold or those consisting of an interconnected particle network, with topology, connectivity and persistence length changing depending on the filler concentration excess over the percolation threshold value.

3.4 "Diffused" and "Fractal" p-n Junctions in Composites Filled with Both p- and n-Type Particles

Even more interesting physics comes into play if both p- and n-type nanoparticles are successfully mixed in the same polymer matrix. Since the percolation threshold for each type of nanoparticles is around 15 vol. %, and can be even smaller due to the hopping conduction, in order to obtain composites with connected infinite clusters of both p- and n-type nanoparticles, one should have 30% of the composite volume occupied by filler and the remaining 70% by polymer. It is possible to synthesize composites with such volume ratios of components and this possibility was realized in our laboratory in 1996 [7]. As described in Sect. 2, depending on the nature of the preparative method, different composite morphologies can be obtained. By co-precipitation from the colloid solution (Sect. 2.1) almost homogeneous mixtures of interconnected nanoparticle networks of both p- and n-type can be obtained, similar to the polymer interpenetrating networks that are currently the subject of intensive investigation [23]. Another peculiar morphology of p-n nanocomposites which can be obtained by means of conversion of p-type particles into n-type ones (Sect. 2.1) is the diffused junction, which is illustrated in Fig. 5. In this case one obtains an area consisting of pure p-type nanoparticles close to one of the electrodes, an area of pure n-type particles near another electrode and an intermixed layer, with the thickness depending on the chemical reaction conditions and consisting of nanoparticles of both types, with changing connectivity of corresponding infinite clusters along the sample thickness.

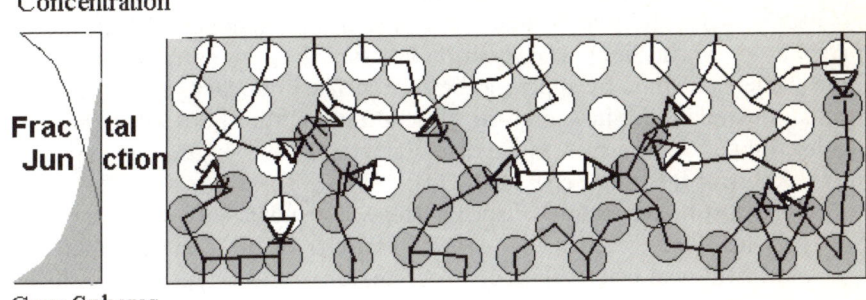

Fig. 5. Structure of a device with fractal junction geometry. L_d-Debye length, R_p-nanoparticle radius, W-intermixed layer width

A number of effects arise in such intermixed layers due to the fact that the size of the nanoparticles is either smaller or comparable to the Debye screening length L_D – the characteristic length to which the field penetrates into the semiconductor:

$$L_D = \sqrt{\frac{\varepsilon\varepsilon_o kT}{e^2 n_b}} \qquad (2)$$

where ε, and ε_o represent the dielectric constants of the material and vacuum, respectively, and n_b is the number of free charge carriers in the conduction band of the particle.

Since it is possible to change the Debye length by changing the concentration of carriers by means of doping or photogeneration, the effect which we termed "fractalization" of the p-n junction is observed: If we have an intermixed layer of nanoparticles with a thickness of ca.7 particle diameters and the Debye length is larger than the particle diameter, the geometry of the space charge layer formed at the p-n junction will be planar, since the field will be insensitive to the microstructure of the layer (or better the nanostructure), and hence be the same as in the case of a planar p-n junction. However, if by means of doping or light irradiation the Debye length can be made comparable to the nanoparticle diameter (50 nm in our case), "fractalization" of the space charge layer takes place, such that the space charge layer now mimics a single p-n nanoparticle junction since the concentration of carriers is high enough to screen the field within one particle diameter. As already known from semiconductor physics, the structure of the double layer at a p-n junction can be obtained by solving the Poisson equation:

$$\nabla^2 \Phi(\mathbf{r}) = 4\pi\rho(\mathbf{r}) \qquad (3)$$

where $\Phi(\mathbf{r})$ is the potential distribution and $\rho(\mathbf{r})$ is the charge distribution. In our case the uncompensated ionized impurity atoms are homogeneously distributed in each nanoparticle, and the p- and n-type nanoparticles are intermixed, as shown in Fig. 3. The equation can be solved numerically using the finite elements method by applying the appropriate boundary conditions. The results of simulations show that the "fractalization" threshold values are dependent on parameters such as particle diameter, Debye length, topology and the thickness of the intermixed layer [24].

Another distinctive feature of such systems is the existence of strong local electric fields. Since, even in the case of particles of the same conductivity type, local fields can reach values of 10^7 V/cm, for p- and n-type nanoparticles in contact with each other the fields will be even higher and, according to our estimations, reach values in excess of 10^8 V/cm.

Thus the space charge level "fractalizes", providing a p-n interface area which can exceed the area of the planar junction by 10–1000 times, depending on the topology of the intermixed region [24]. This is especially advantageous for the manufacture of photovoltaic nanocomposite cells, since the active area, i.e. the area where local fields are strong enough to separate the photogenerated excitons, created in both the polymer and in the nanoparticles is increased up to values of 10–20 microns. The topology of the fractal junction also provides continuous pathways for charge carriers of both types to the corresponding electrodes with minimized recombination, since in the proximity there is always a particle of the appropriate conduction type connected to the percolative network to which it is possible to jump. In this way holes can be transferred through the connected network of p-type particles and electrons through the connected network of n-type particles, co-existing in the space of the active area of the diffused junction with given topology.

Other possible applications of gradient junctions are in polymer-nanocomposite LEDs, where a voltage-dependent color is expected, and even for the manufacture of lasers with electrical pumping, since the injection current densities in the case of gradient junctions can be made quite high. However, in the case of lasers, the nanoparticles must be fairly uniform in size, a condition which is barely achievable by the chemical synthetic methods usually used for the preparation of nanocomposites.

Physics similar to those described above are responsible for the operation of interpenetrating networks based photodiodes and the solar cells based on polymer-fullerene donor-acceptor transitions which are actively under investigation at present [25]. Since fullerenes in such blend-based devices are in the nanocrystalline form, with cluster sizes of the order of 2 to 10 nm, they are also polymer-nanocomposites and hence similar to the other objects treated in this review. The device geometry of the fullerene-polymer solar cells is based on the so-called bulk heterojunctions, which are similar to the gradient junctions described above. The distinctive feature of polymer-fullerene blends is the low concentration of majority carriers especially without illumination, which makes it impossible to form a space charge layer between p- and n-type areas. Thus the mechanism of their operation is much closer to the molecular concepts of photoinduced charge transfer and Onsager charge separation in the external field, provided by the difference in work function of two electrodes, rather than to traditional semiconductor p-n junction principles [26].

4
Electroluminescence in Polymer-Nanocomposites and Light Emitting Diodes

Electroluminescence is an electronic analogue of photoluminescence and consists of the radiative recombination of the electrons and holes injected into the conduction and the valence band of the semiconductor, respectively (Fig. 6). Electroluminescence is intensively exploited in optoelectronic devices based on inorganic semiconductors such as light emitting diodes (LEDs) and lasers.

The important criteria which the material must fulfill to be suitable for the manufacture of electroluminescence devices are high injection current densities of both the electrons and the holes into the active region of the device, a

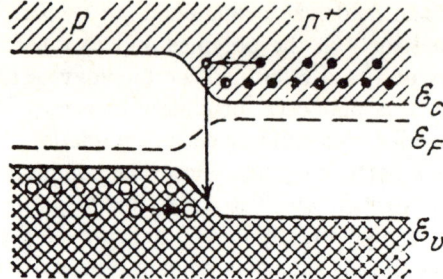

Fig. 6. Scheme of radiative recombination in a p-n junction. *Left area* base; *right area* emitter

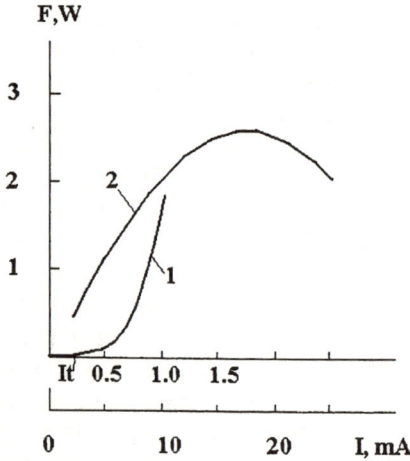

Fig. 7. Radiative characteristic of LEDs (dependence of emitted radiation flux F on current I). *1* low current branch; *2* high current branch

much higher probability of radiative recombination of electron-hole pairs in the active device area in comparison with the non-radiative channels and an optical transparency of material in the emission spectral range.

The most important device parameter reflecting the electroluminescent devices performance is the external quantum efficiency (EQE), which is determined by the expression:

$$\eta = \gamma \eta_e \eta_{opt} , \qquad (4)$$

where γ is the injection coefficient, η_e-the internal quantum efficiency and η_{opt}-the light output coefficient. The most informative are the peak or integral brightness-current dependencies (Fig. 7) and the spectral content of the emitted radiation. EQE for the best inorganic semiconductor LEDs reaches values of 20–30%. Other important parameters of an LED are the integral

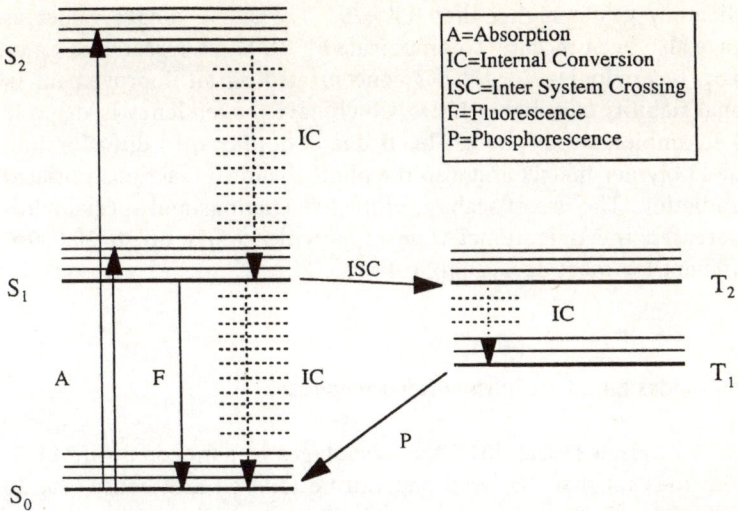

Fig. 8. Different processes which occur in a conjugated polymer molecule on photoexitation

brightness or peak brightness, i.e. the amount of radiation emitted over the whole spectral range or at the peak wavelength, as well as the threshold voltage, i.e. the voltage at which radiative emission begins.

Recently, organic polymer materials have been applied in electroluminescent devices such as LEDs, primarily for display applications, and as optically pumped lasers [27-29]. In the case of polymer-based structures, a molecular picture of electroluminescence is more adequate than the semiconductor band scheme (Fig. 7), i.e. radiative transitions between the electronic states of excited conjugated molecules and the ground state take place. It was discovered that before their recombination in the conjugated polymer media the electrons and holes succeed in forming excitons (bound states of the electron-hole pair) which than recombine through a number of radiative and non-radiative channels (Fig. 8). The polymer used as the electroluminescent media must combine high electron and hole injection-based mobility with a high probability of radiative recombination in comparison with the non-radiative channels.

One advantage of polymers which is especially useful for laser-oriented applications is the red shift in the luminescence spectrum in comparison with their absorption. This phenomenon gives the opportunity of light amplification without losses caused by self-absorption=in active media. Many efforts are now being concentrated on the manufacture of electronically pumped polymer lasers, although they have not yet met with success.

Using molecular engineering approaches, physicists and chemists have successfully optimized the structure of conjugated polymers, mostly poly(phenylenevinylene) (PPV) or polythiophene (PT) derivatives, for LED applications to obtain all the colors of the visible spectrum and recently have reached quantum efficiency values higher than 1% [29]. Threshold voltage values as low as 1 V have also been recently communicated [30]. Much research at Cambridge Display Technologies Co. [1] has concentrated on an improvement in the operational stability of polymer LEDs, which have the tendency to degrade quite fast in an ambient atmosphere. This is due to both oxygen diffusion into the conjugated polymer matrix and also the photochemical reactions initiated by emitted radiation. The use of sealing, protective coatings and special additives has increased the operational time of polymer LEDs up to 20,000 h, which is sufficient for most device applications.

4.1
Light Emitting Diodes Based on Polymer-Nanocomposites

According to P. Alivisatos et al. [31], the advantages of nanocomposite LEDs originate from the fact that the band gap can be tailored and hence, due to quantum confinement, the emission color changes depending on the size of the nanoparticles (see Sect. 2). This phenomenon is well illustrated in Fig. 9, which displays the effects of quantum confinement in photoluminescence, electroluminescence and optical absorption spectra.

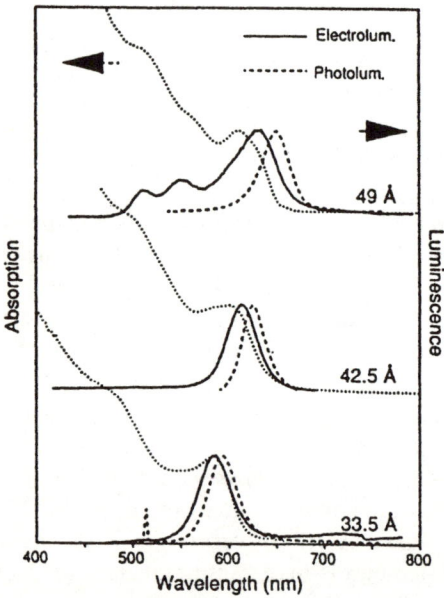

Fig. 9. Quantum confinement effect [22] in the electroluminescence (*solid line*), photoluminescence (*dashed line*) and absorption (*dotted line*) of nanocrystals of three different diameters. The blue shift of the (49, 42.5 and 33.5 Å) electroluminescence in respect to the photoluminescence may be due to trap filling by carriers injected into the CdSe multilayer or to a charge transport mechanism in the CdSe multilayer

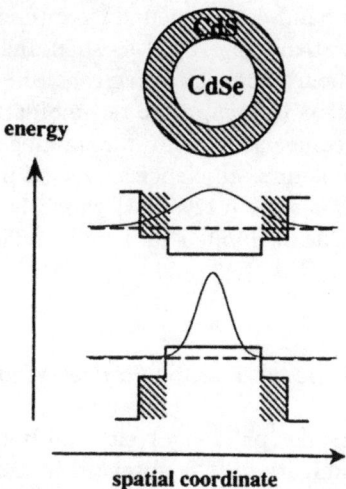

Fig. 10. CdSe(CdS) core-shell nanocrystal [24] energy diagram. The *solid lines* represent the bulk potentials and the *dashed levels* represent the HOMO and LUMO in the nanocrystal extrapolated from bulk CdSe. The Gaussians shown are the results of effective mass calculation for electron and hole wave function. (see also [40]). Potential offsets between the core and shell valence band (0.5 eV) and conduction band (0.3 eV) are approximated by the differences between the bulk electron affinities and ionization potentials of CdSe and CdS

Another striking feature of polymer-nanocomposite LEDs is the voltage-dependent color of the emitted light. The phenomenon is caused by recombination in polymer-nanocomposite systems which usually takes place both in the nanoparticles and in the polymer matrix. Varying the ratio of electron-hole pairs undergoing recombination in the matrix and the nanoparticles correspondingly changes the injection conditions from the cathode and anode, which affect the range of carriers penetration within the structure resulting in a variation in the spectral content, i.e. the color of the emitted light. Since electron injection and transport are usually limiting for polymer and polymer-nanocomposite devices [31], it was pointed out that changes in the electron injection and the electron penetration depth caused by the change in voltage are responsible for the shift in recombination zones which results in the emission color change. Even though similar voltage-dependent color LEDs have been reported based on electroluminescent conjugated polymer blends consisting of polymers with different band gaps in an insulating polymethylmethacrylate (PMMA) matrix [32], polymer-nanocomposites seem to be easier to process and produce the desirable color coordinates changing the particles sizes and the thickness of the layers.

A further conceptual development of polymer-nanocomposite LEDs, reported by Alivisatos et al. [33], was the application of core-shell CdSe-CdS structures. Due to the specificity of the confinement potential (Fig. 10) caused by the fact that the lower band gap semiconductor (CdSe) forms the core of the particle and the higher band gap (CdS) its shell, the holes are trapped in the core area of the nanoparticles while electrons are delocalized over the whole nanocomposite system. Therefore the recombination probability is enhanced significantly in comparison with the homogeneous nanoparticles layers. The best external quantum efficiencies reached using core-shell CdSe-CdS nanocomposite systems were 0.22% [34] which is at least one order of magnitude larger than values exhibited by simple CdSe-based nanocomposites.

5
Photoconductivity and Photovoltaic Effect in Polymer-Nanocomposites

Photoconductivity reflects the process of electron-hole pair generation in semiconductors upon illumination. It is observed in the form of an increase in current if the semiconductor is irradiated by a flux of photons with an energy greater than the semiconductor band gap value. This phenomenon is utilized in devices such as photoreceptors and photodetectors of various kinds, which convert the optical signal into an electrical current. However, for the purpose of photodetection, photodiodes, i.e. devices utilizing p-n junctions or Shottky metal-semiconductor contacts, are used rather than photoresistors. In photodiodes the electric field at the p-n junction or at the interface between the metal and the semiconductor separates the photogenerated electron-hole pairs preventing their recombination and thus increasing both the photocurrent and the light sensitivity of the photodiodes in comparison with photoresistors. Usually both p-n junction diodes and Shottky diodes are used in the reverse bias regime (Quadrant III in Fig. 11) maximizing the energy barrier at

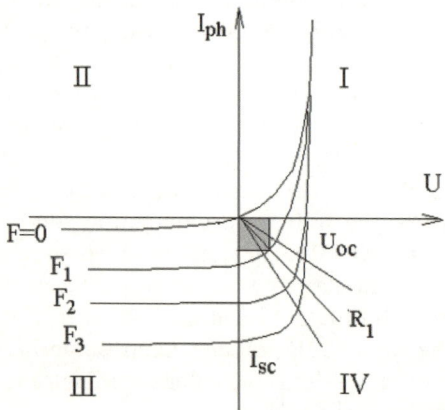

Fig. 11. Set of curves (I-V) of the photodiode for different illumination flux intensities Φ. *Quadrant I*-forward bias area (not used in optoelectronic devices); *Quadrant III*-photodiode operation area; *Quadrant IV*-photovoltaic operation area

the junction, which becomes almost non-transparent for the diffusion-originating current of the majority carriers (i.e. holes in the p-type area and electrons in the n-type area). At the same time the photogenerated minority carriers see no barrier at all and are collected by the corresponding electrodes. Thus a high ratio of photocurrent to dark current and hence photodiode sensitivity is reached, which is desirable for photoreceptor and photodetector applications.

One of the main parameters of a photodetector is the spectral current sensitivity (I_f), expressed by the following formula:

$$I_f = q\eta\gamma \frac{F}{h\nu} \, , \tag{5}$$

where h is the internal quantum yield, i.e. the number of electron-hole pairs produced by one photon with given frequency ν, γ_x-is the transfer coefficient, which reflects the ratio of the carriers which do not undergo recombination and are transferred to electrodes to the total number of carriers, and F is the light intensity. Figure 12 shows the corresponding dependence of the current on the integral photon flux which is one of the other vital parameters of a photodiode. Another important characteristic of a photodiode is the rectification ratio in the dark and under illumination, which is the ratio between the forward bias current and the reverse bias current at a selected voltage, which must be high enough to lower the barrier in the case of reverse bias and to elevate it in the case of forward bias. The rectification ratio reflects how "good" the photodiode actually is as a current rectifying device.

The photovoltaic effect, which is utilized in the manufacture of solar cells, corresponds to quadrant IV of the current-voltage (I-V) curve drawn in

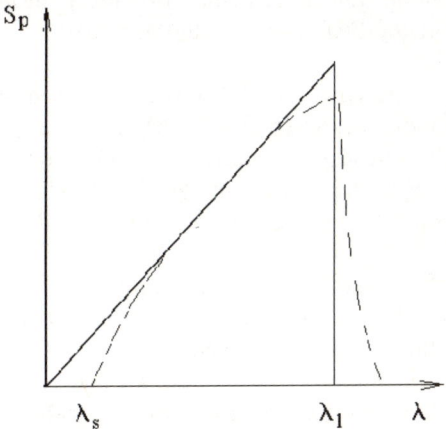

Fig. 12. Typical spectral response function of a photodiode. The long wavelength edge of the spectrum λ_l is determined by the band gap, the short wavelength limit λ_s is determined by the increase in the absorbance in the passive layers of the device once the wavelength is decreased

Fig. 11. In this operational mode there is no bias voltage applied to the device and the separation of photogenerated electrons and holes at the p-n junction provides the electromotive force, i.e. the stable potential difference at the electrodes which can be kept under resistive load, which means that a photovoltaic cell is an electric power source. The main electrical photovoltaic parameters of the solar cell are the open circuit voltage (V_{oc}), i.e. the voltage under disconnected electrical circuit conditions, the short circuit current (I_{sc}), i.e. the current which can be obtained by short contacting the device to itself, and the fill factor, which is determined by following expression:

$$FF = \frac{I_m U_m}{I_{sc} U_{oc}} \qquad (6)$$

where U_m and I_m are the values of the voltage and current, respectively, which provide the maximum power output of the cell.

The most important characteristic of a photovoltaic solar cell is the energy conversion efficiency, which is determined by the equation:

$$\eta = \frac{I_{sc} U_{oc} FF}{P} \qquad (7)$$

where P is the integral radiation power per device surface area.

Common commercial inorganic solar cells based on polycrystalline silicon usually possess efficiency values 10 to 12% under solar illumination (AM 1.5 1SuN).

Polymer photodiodes have recently attracted the interest of materials scientists and technologists due to the opportunity to make large area arrays of addressable photodiodes [34] which could provide a cheap alternative to vidicons and MIS-capacitor CCD used in digital video cameras or night vision devices.

On the subject of polymer or rather organic-based solar cells, these attracted the attention of researchers in the early 1970s as quite promising cheap and environmentally friendly alternatives to their inorganic prototypes. The best results obtained so far are the molecular crystal p-n junction devices made of perylene and copper phthalocyanine [35], and the energy efficiencies reached were of the order of 2%.

Quite recently efforts have concentrated on the development of polymer and polymer-nanocomposite plastic solar cells [36]. The main advantage of polymer-based cells is the possibility to make large area flexible structures using inexpensive processing methods such as doctor blading or solution casting.

In fact, prior to the manufacture of photovoltaic cells, one always examined the extent of photoluminescence quenching in the materials, which gave a good marker of how well the electron-hole pairs had been separated either by the internal field (Onsager dissociation), or as a result of other processes such as photoinduced charge transfer.

5.1
Photodiodes and Solar Cells Based on Polymer-Nanocomposite

Returning to polymer-nanocomposites, we have studied the photoluminescence quenching in nanocomposites based on CdS nanoparticles with different sizes in a poly(vinyl alcohol) (PVA) matrix (Fig. 13), which is non-conductive, while a group at Berkeley University has concentrated on substituted poly(paraphenylenevinylene) (MEH-PPV)-CdSe nanocomposites, MEH-PPV being a conductive as well as a luminescent matrix in its own right (Fig. 14).

In both cases one can see that the luminescence is quenched once the concentration of nanocrystals in the matrix is increased. Quenching in the case of the PVA matrix is more complete, since PVA is a non-luminescent matrix and thus only the luminescence of CdS nanocrystals must be quenched. Even though the two systems seem to be similar, the processes taking place are different since in the case of PPV-CdS systems the electron-hole pairs are generated both in the nanoparticles and in the polymer matrix; electrons are then transferred through the nanoparticles network, the holes tend to be transported through the polymer. In the case of CdS nanoparticles in the non-conductive and non-luminescent PVA matrix the electron-hole separation takes place between neighboring nanocrystals, assisted by the high local fields as described in Sect. 3, and hence the matrix plays a passive role in the photocurrent generation process.

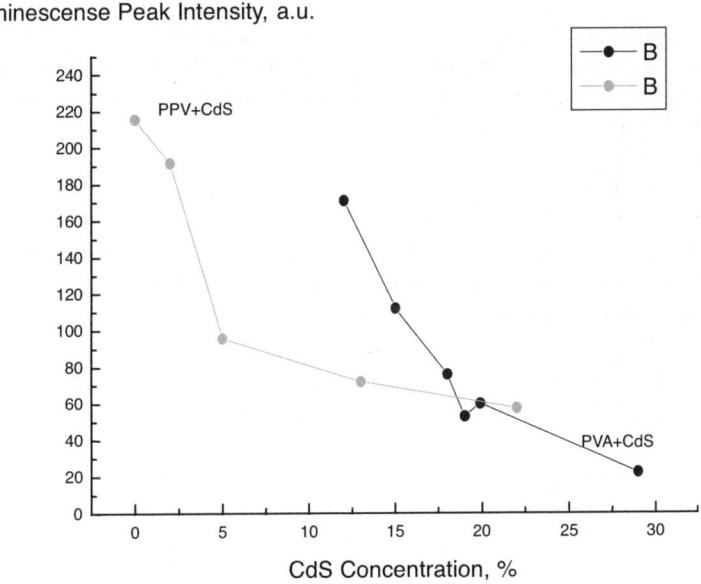

Fig. 13. Height of luminescence peak [13] for CdS-PVA nanocomposite (**a**) and CdS-PPV (**b**) on CdS concentration. Nanoparticles size increases from 2 nm at low concentrations to ca. 4 nm at higher ones

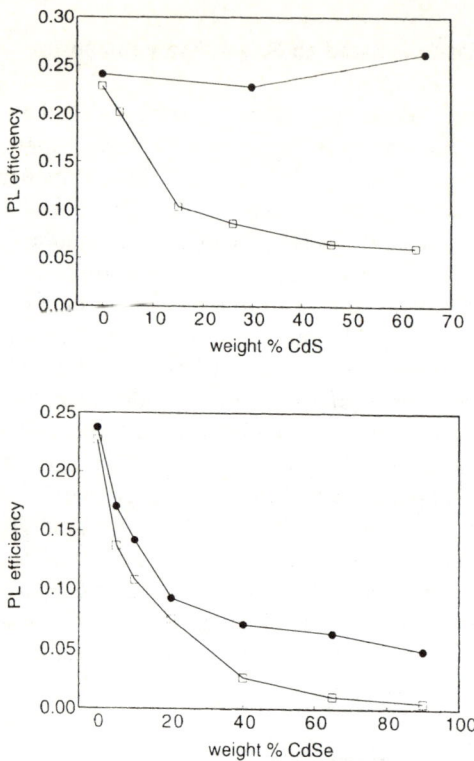

Fig. 14. Photoluminescence efficiency [29] of MEH-PPV/CdSe and CdS nanocomposites as a function of nanocrystal concentration, for TOPO-coated (*circles*) and pyridine-treated (*squares*) nanocrystals. **a** 4-nm diameter CdS nanocrystals; **b** 5-nm diameter CdSe nanocrystals

Let us now describe the photovoltaic properties of the polymer-nanocomposites that were mentioned above. A number of structures which have been studied are depicted on Fig. 15. We regard the nanocomposites containing highly doped p- and n-type nanoparticles to be more promising for photovoltaic applications than ones filled with particles of one conductivity type only for the following reasons:

1. All the advantages of fractal p-n junction can be utilized, i.e. producing photovoltaic cells with an optimal thickness of the active layer of ca. 10 microns.
2. All the advantages of heterojunction II-VI sulfide-based solar cells can be utilized, with the polymer matrix preventing interdiffusion of A_{II} metal components, often the main reason for the instability observed in inorganic heterostructure II-IV solar cells produced by liquid epitaxy or thermal evaporation.
3. The conductive polymer matrix also plays the role of an i-type semiconductor, thus allowing p-i-n device structures, with extremely high local fields

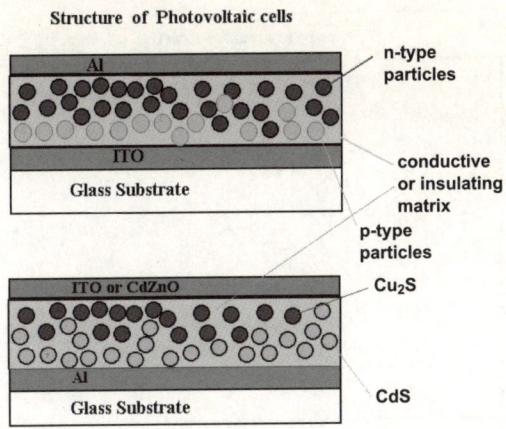

Fig. 15. Structure of p-n nanojunction based photovoltaic cells

in the i-area, causing decomposition of excitons generated in the polymer matrix. The latter fact leads to the additional possibility of photon harvesting, tailoring the matrix absorbance band in order to increase the total energy conversion.

4 The topology of conduction networks formed by p- and n-type particles provides a gradual increase in the connectivity in the built-in field direction, which minimizes the amount of network dead ends in the nanoparticle chains, at which the charge carriers are usually trapped, and which are the main centers of recombination in polymer-nanocomposites.

5 The p-n nanocomposite photovoltaic cells can be made with a large area and be flexible.

We investigated a system of p-type Cu_2S nanoparticles mixed with n-type CdS, doped with indium in order to increase carriers concentration, synthesized in situ in a poly(vinyl alcohol) matrix. Cu_2S and CdS placed in contact form a heterojunction which delivers 12% of the power conversion efficiency when made by liquid epitaxy. Nevertheless, the inorganic CdS/Cu_2S cells produced by epitaxy suffer from a number of disadvantages, one of which is the degradation caused by diffusion of Cu^+ ions into the CdS layer. In our case, by preparing the nanocomposites of CdS and Cu_2S in a polymer matrix, we prevented diffusion, thus improving the long-term stability of the photovoltaic cells significantly.

Energy efficiencies in the order of 2% under solar illumination at AM0 1 sun conditions were obtained on the systems described above [6], which is much lower than expected, and we continue to channel efforts in this direction with the aim of reaching efficiency values of the order of 10%.

Another type of devices are the photovoltaic cells [37, 38] formed by n-semiconductor nanoparticles impregnated into a conductive p-type matrix

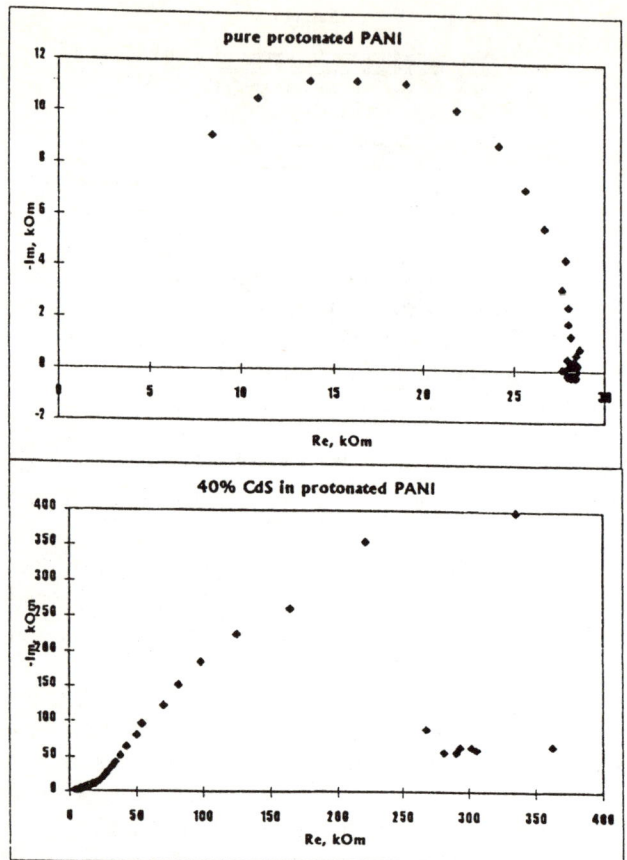

Fig. 16. Impedance spectra (plot of the imaginary part of complex impedance and real on frequency) for **a** Ag/PANI/Ag and **b** Ag/PANI+30%CdS/Ag

(Fig. 15). CdS and CdSe nanoparticles have chiefly been used as the filler and MEH-PPV, p-PPV and polyaniline as the host matrices.

We have demonstrated [37] that p-n nanojunctions are formed between the nanoparticle chains and the polymer matrix with the partial formation of a space charge layer. The evidence is a 3 orders of magnitude drop in the composite conductivity, a change in the impedance behavior (Fig. 16) and a difference in the I-V characteristics (Fig. 17). The most probable cause is the formation of a space change layer which, in turn, generates local electric fields, leading to a decrease in concentration of holes, and a mobility drop due to an increase in the disorder.

A number of processes leading to electron-hole pair separation take place in polymer-nanocomposites [38], as seen in Fig. 18. Charge transfer mainly takes place at the particle-polymer interface and the photocurrent spectrum

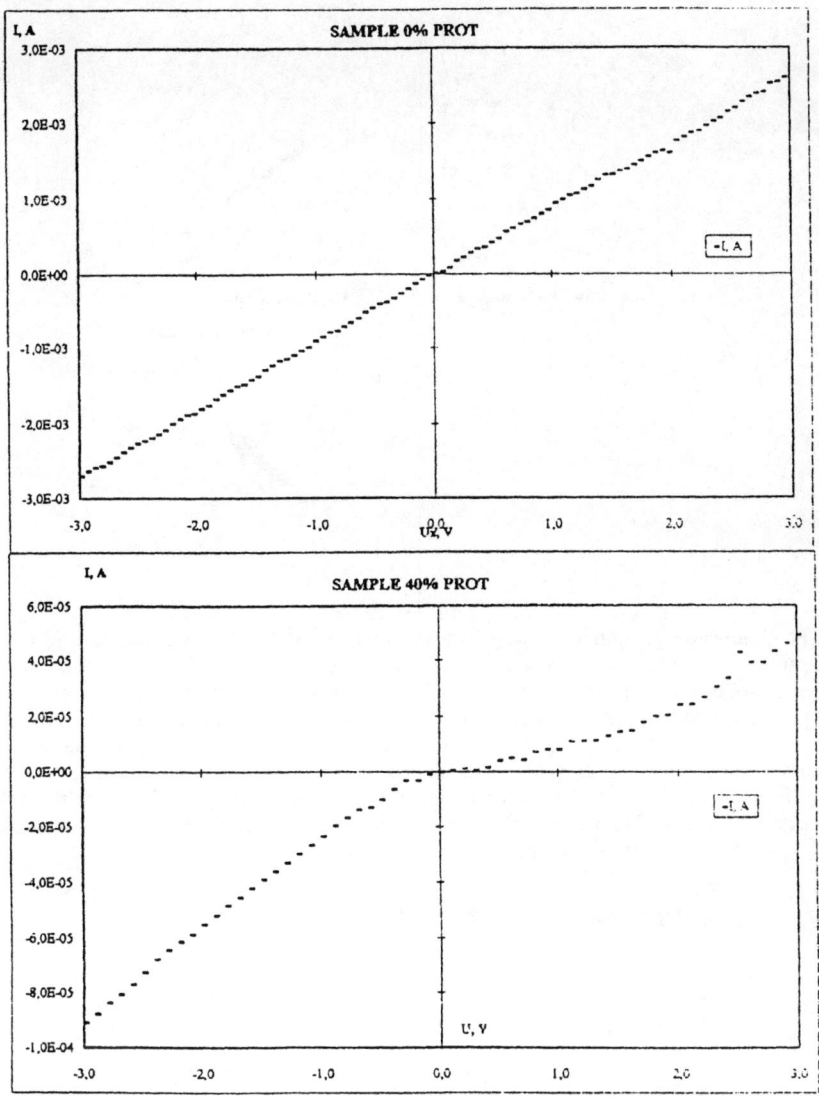

Fig. 17. I-V curves for **a** Ag/PANI/Ag and **b** Ag/PANI+30% CdS/Ag

(Fig. 19) provides evidence that a photocurrent is generated if the photon is absorbed by both the nanoparticles and the polymer matrix.

If the nanoparticle surface is coated by a capping organic molecular layer only an exciton transfer is possible [38], since the charge transfer is suppressed due to the presence of the relatively thick insulating layer. The usual case is that wherever the exciton is generated, the hole tends to come to rest and is transferred to the anode through the polymer matrix, while the electron

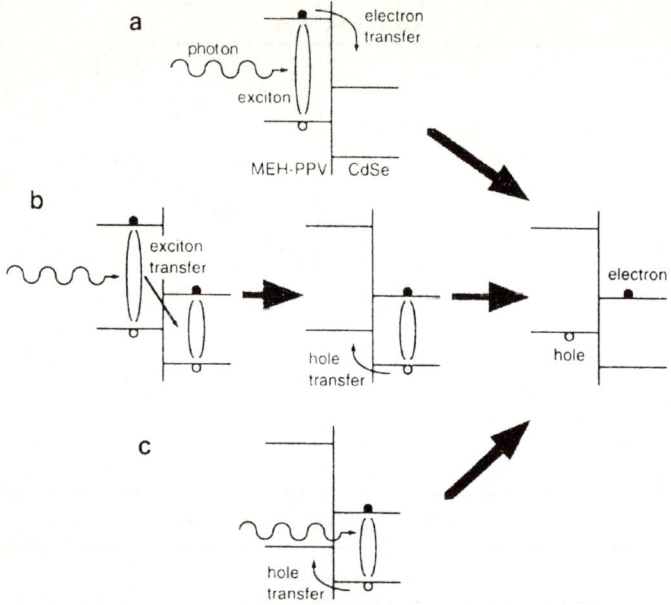

Fig. 18. Routes for exciton and charge transfer [29] in MEH-PPV/CdSe blends. *a* Absorption in the polymer, followed by electron transfer to the nanocrystal. *b* Absorption in the polymer, followed by exciton transfer to the nanocrystal, followed by hole transfer onto the polymer. *c* Absorption in the nanocrystal, followed by hole transfer onto the polymer. Note that for CdS nanocrystals, route *b* is not available since the nanocrystal energy gap is larger than that of the polymer, whereas it would be possible for excitons to transfer from the nanocrystal to the polymer, followed by electron transfer onto the polymer. In the presence of a TOPO barrier, the electron and hole transfer processes are suppressed and only exciton transfer is possible

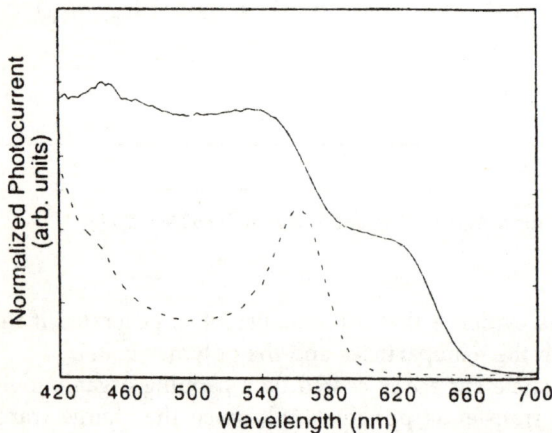

Fig. 19. Spectral response [29] of the short circuit current for a MEH-PPV-CdSe device containing 90 vol% 5-nm diameter CdSe nanocrystals (*solid line*) and for a pure MEH-PPV device (*dashed line*). The data have been normalized to fit on the same scale

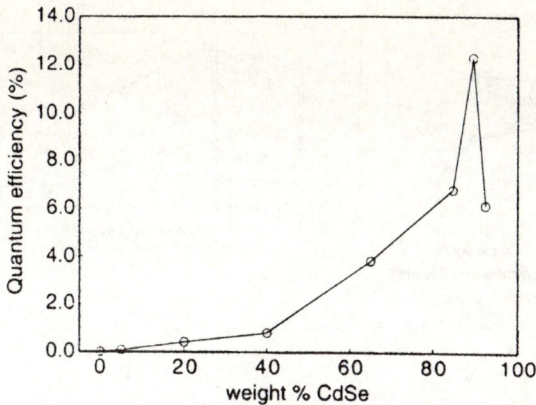

Fig. 20. Short circuit quantum efficiency [29] for devices containing 5-nm diameter CdSe nanocrystals as a function of CdSe concentration. Excitation: 514 nm, Power density: 5 W/m^2

is transported through the connected chains of the CdS or CdSe nanoparticles. It was found that, at high filler concentrations, up to 90% of all generated excitons are decomposed, and thus the transport to the electrodes is the main factor limiting the photocurrent. Efficient transport requires high concentrations of the nanoparticles and Fig. 20, which depicts the short circuit quantum efficiency versus filler concentration, illustrates this fact. The main losses are most probably due to the trapping of electrons at the dead ends of the conduction network, which act as the recombination centers [38].

The best photovoltaic properties were exhibited by a structure having 90 wt.% of CdSe. The energy conversion efficiency when illuminated by the solar spectrum was found to be around 0.1%, which is not as high as the values exhibited by inorganic solar cells. However, the system can be significantly improved by adding dye molecules to the polymer matrix and optimizing the topology of the conductive nanoparticles networks using the gradient approach described in Sect. 2.

The author with co-workers [39] has investigated systems with the structures Al/C$_{60}$/PANI+CdS/ITO and Al/C$_{60}$/PPV+CdInS/ITO. Most worthy of note are the dependencies of the short circuit current and the open circuit voltage on the nanoparticles concentration (Fig. 21). Since the fullerene molecule acts as a strong electron acceptor, excitons generated both in the polymer matrix and in the CdS particles are decomposed, electrons are accepted by the C$_{60}$ layer and the holes are transported to the anode through the polymer. Once the concentration of the nanoparticles exceeds the percolation threshold value, the system becomes short contacted, since the electrons can pass from the anode to the cathode through the barrier-free connected network of CdS clusters; this latter fact leads to the disappearance of the photovoltaic effect, as illustrated in Fig. 21b. The increase in the photocurrent and the open circuit

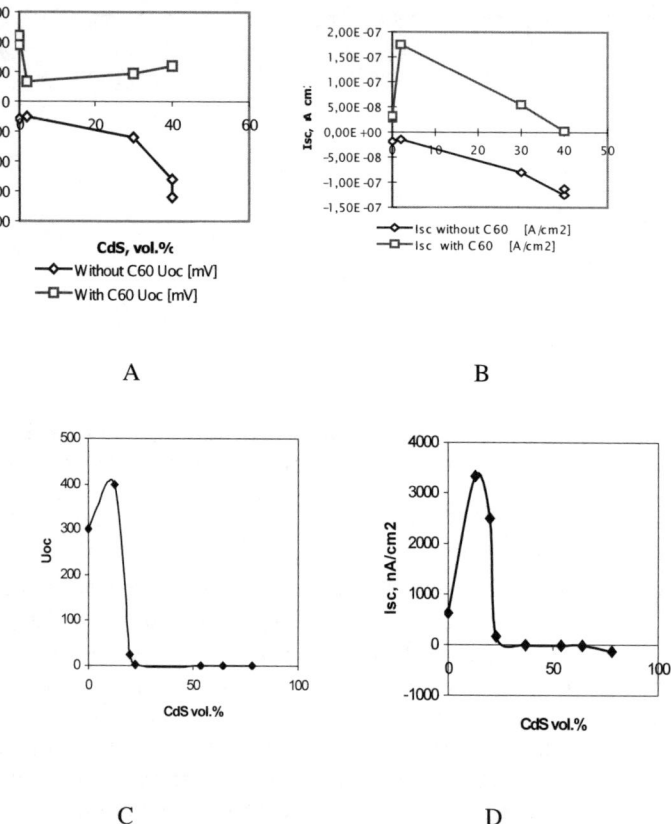

Fig. 21. Short circuit current (I_{sc}) (**A, C**) and open circuit voltage (V_{oc}) (**B, D**) dependencies on CdS concentration for Al/CdS+PANI/CdZnO (**A, B**) and Al/CdS+PPV/CdZnO (**C, D**) with and without a fullerene layer between Al and the composite

voltage with the increase in concentration of the nanoparticles preceding the percolation threshold is most probably due to the additional exciton generation in CdS nanoparticles.

There is also some evidence that the chains of CdS nanoparticles possibly play the role of the anode (hole collector) dispersed into the polymer matrix. If the valence band levels of the polymer matrix and the nanoparticles match each other well, as in the case of the PANI/CdS system, a hole can be injected into the nanoparticle and recombine with the intrinsic electron, charging the particle positively. The positive charge can then be transferred to the anode since the hopping probability to a positively charged particle is larger than to a neutral one (Fig. 22):

$$R_{ij} = R_{ij}^o \exp\left\{-\frac{r_{ij}}{\beta}\right\} - E_{ij} \tag{8}$$

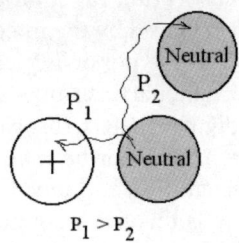

Fig. 22. Diagram of the hopping conduction in a CdS/PPV nanocomposite layer

$$E_{ij} = \frac{q^2}{\varepsilon R k T} \tag{9}$$

$$E_{ij} = \frac{q^2}{2\varepsilon R k T} \tag{10}$$

where R_{ij} is the resistance of the equivalent network element, which corresponds to the hopping in between an i- and a j-nanoparticle, $R^o{<}{?}{>}_{ij}$ is the pre-exponential factor, r_{ij} the distance between nanoparticles, E_{ij} the energetic part of the hopping process, β the effective radius of the eigenstate and ε the dielectric constant of the media.

The electrostatic energy term in the case of hopping to a neutral nanoparticle (Eq. 9) is two times bigger than the term reflecting the hopping to a positively charged one (Eq. 10). Thus the equivalent network resistance (Eq. 8) (using the Miller-Abrahams approach) becomes smaller giving the possibility of a transfer of a positive charge to the anode not in a form of a hole but as a positive charge delocalized over the whole nanoparticle. This process, if it exists, provides the explanation for the increase and following drop of both U_{oc} and I_{sc} which take place for the filler concentration values in the vicinity of the percolation threshold for CdS-p-PPV nanocomposites (Fig. 21b).

6
Gas and Vapor Polymer-Nanocomposite Sensors

The possibility of utilizing polymer-nanocomposites for the manufacture of gas and vapor sensors was studied in our laboratory in 1992 [40]. The specificity of nanocomposites, which makes them attractive as a gas sensor material, is the existence of a hopping conductivity between the nanoparticles through the polymer in a range of concentrations close to the percolation threshold (see Sect. 2.3). Since the absorption and partial dissolution of the gas or vapor molecules in polymer matrix changes its properties, it is possible to monitor these changes measuring electrophysical characteristics of the composites such as conductance and capacitance. The important point here is the possibility to separate the absorption function and the sensor transducer function of the device: the absorption function is played by the polymer matrix and the sensor function is played by the interconnected nanoparticles network. This separation provides the possibility of combining various polymer matrices with different filler nanoparticles thus allowing selective detection of some gas molecules if the polymer matrix absorbs them selectively. By varying the polymer matrix, sensor arrays can be manufactured and these are now extensively used for odor recognition and in various biosensor applications.

Another distinctive feature is the very magnitude of the effect: the conductivity change in the threshold area is usually 6–7 orders of magnitude, thus making the conductivity change upon absorption of the gas molecules large as well.

Two basic factors determine the conductance and capacitance change during gas or vapor absorption: the adsorbate swells the polymer matrix and the dielectric properties of the polymer matrix are changed. Swelling leads to a change in interparticle distances, while the change in dielectric properties influences hopping conduction. Both of these effects are utilized for sensor operation, though swelling is a slow process, and the corresponding sensor response times are tens of minutes, being insufficient for most applications.

The change in the dielectric properties of the polymer matrix, mainly of the dielectric constant, leads to a dramatic change in conductance due to the peculiarities of the hopping mechanism. In the course of our research [41] we treated the hopping conduction in polymer-nanocomposites in the framework of the Shklovski-Efros model of hopping conduction in semiconductors. The basic equation which determinies the composite conductivity in this case is:

$$S = S_o \exp\left(\frac{-2r_{ij}}{\beta} - \frac{q^2}{\varepsilon kT} + F\right) \qquad (11)$$

where S_o is the proportionality factor, r_{ij} is the interparticle distance, β-the characteristic radius of electron eigenfunction delocalization, ε-the dielectric permeability of the media and F defines the term responsible for the electric field influence on hopping due to the potential barrier lowering.

The first term under the exponent reflects the dependence of the hopping probability on the distance between the nanoparticles while the second term handles the energetic part of the hopping process, the energy being mainly of electrostatic origin. Since both the first term (through β) and the second one include the dielectric permeability of the media, the hopping conduction exponentially strong depends on the value of ε. This means that even small changes in polymer matrix dielectric permeability caused by absorption of polar molecules lead to an exponential change in the composite conductivity. This process is much faster than a swelling-induced conduction change, since it requires only the adsorbate molecule penetration into the polymer in between the nanoparticles and no chemisorption or dissolution of the molecule in the matrix is necessary.

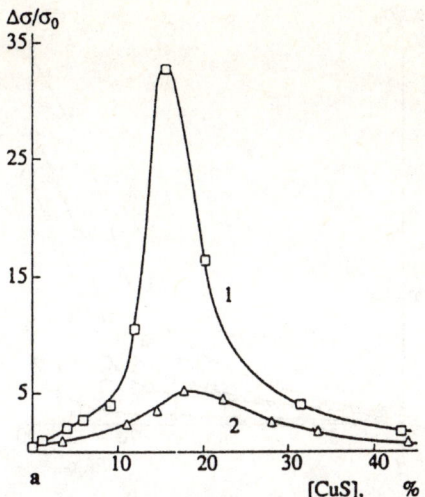

Fig. 23. Dependence of the relative change in CuS-PVA composite layer conductance on the increase in relative humidity from 15 to 74% RH on CuS content (vol%). *1*-Nanocomposites with particle size ca. 12 nm; *2*-Micrometer size particles (10–15 mm)

The first sensors to be manufactured utilizing the principles described above were humidity sensors, since a water molecule possesses a large dipole moment and thus changes the dielectric permeability upon absorption to a great extent. We studied CuS-PVA nanocomposites as prospective humidity sensors and measured the dependence of the relative conductance change on the the nanoparticles concentration (Fig. 23). It can be seen that for the nanocomposites with a filler concentration within the range 10–15 vol%, i.e. just preceding the percolation threshold, conductivity changes 35 times when the relative humidity (RH) is increased from 15 to 73%. Figure 24 shows the dependence of the relative conductance change on the partial pressure of water vapors (relative humidity). It can be seen that the dependence of conductivity on RH is the exponential one, in full accordance with Eq. (11) if we assume that absorption follows the Langmuir isotherm.

Typical sensor response times were in the order of 5 s, thus excluding the polymer matrix swelling as a reason for the sensor signal. The conductivity change was not particularly large in the described case since the nanocomposite films used as the sensors were quite thick (2 microns), and the bulk conductivity, which is not affected by adsorbate molecules, shunted the surface conductance variation caused by adsorption.

The next step, performed by Trakhtenberg et al. [42], was the use for sensor applications of materials that consisted of thin polymer-nanocomposite layers of poly-*p*-xylylene with lead particles as a filler. The composites were produced by gas phase co-deposition on a quartz substrate, as described in Sect. 2.7. The films made by this method were used to detect ammonia in air

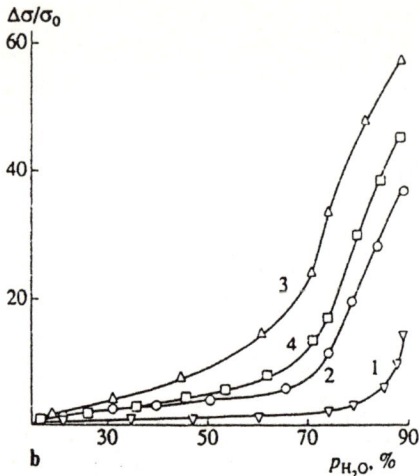

Fig. 24. Dependence of the relative change of nanocomposites conductivity on relative humidity for nanocomposites with filler size ca.12 nm and different concentrations: *1* 7 vol %, *2* 11 vol %, *3* 17.5 vol %, *4* 22 vol %

[43]. The introduction of ammonia impurities changed the sensor conductivity by 4–8 orders of magnitude due to the absence of bulk shunt resistance. The conductivity was found to be proportional to the square of ammonia impurity concentration and the I-V curves of the films were logarithmic, which is typical for the hopping conduction mechanism. By changing the preparative conditions used to produce the films, sensors for the detection of ethanol and water vapors in air were also manufactured.

Composite films of palladium nanoparticles in poly-*p*-xylylene, produced by the same method, were found to be sensitive to small (down to 10 ppm) hydrogen impurities in air. Sensors based on such films can be operated at room temperature [44].

The capacitance change upon absorption in nanocomposites with the filler concentration lower than the percolation threshold is another opportunity to monitor the dielectric permeability change. The effect consists of the existence of the dielectric constant anomaly for the conductive filler concentrations just below the percolation threshold values. The anomaly is due to the giant cross capacitance of big clusters of nanoparticles formed close to the percolation threshold which are not yet in contact with each other (the effect is similar to other critical phenomena, e.g. the behavior of ferromagnetics close to the Curie temperature). The absorption of the gas molecules by the polymer matrix leads to a highly non-linear capacitance change. The use of such polymer-nanocomposite layers placed on the top of field effect transistors (ChemFETs) has been proposed [41], but not yet realized.

7
Polymer-Nanocomposites as Magnetic Storage Materials: Maximum Storage Density Problem

Even though the magnetic properties of nanocomposites have been studied since the early 1960s, interest in nanosize magnetic media has grown only recently due to the development of nanotechnology and nanoelectronics, as well as because of the research in the field of spin electronics devices such as magnetic random access memory (MRAM). Another reason is the quest to increase magnetic media storage capacity. Currently, polymer-nanocomposites are being extensively studied as potential high capacity magnetic storage media and as possible elements of nanoscale spirtronics.

We will describe here the behavior of ferromagnetic iron oxide nanoparticles in a PVA matrix synthesized by the in situ method [45]: Reaction between the immobilized particles in the volume of polymer iron ions (which are attached by means of coordinating bonds to the PVA hydroxyl groups) and the OH^- groups of NaOH followed by treatment in a reducing atmosphere leads to the formation of mixed iron oxide (Fe_2O_3/FeO) ferromagnetic nanoparticles.

The driving force behind these investigations is the problem of maximum density of magnetic storage. In simple terms it is formulated as follows: up to what limit can we increase the density of magnetic storage without losing the reliability of the stored information?

Since the iron oxide particles in a polymer matrix are single domain ones, it is a good model system to study the problem since it is easy to investigate the influence of interparticle distance on magnetic properties by changing the concentration of particles in the polymer matrix, the latter playing the passive role in this case. Another parameter which can be varied is the anisotropy of the particles, which determines the magnetic anisotropy energy. The anisotropy of the particles can be controlled by growing particles in uniaxially stretched polymer matrices. Since the size of the particles is less than the average domain size (ca.50 nm for Fe or γ-Fe_2O_3) all the nanoparticles are in the single domain state, with the magnetic momentum of the particle being the sum of moments of constituent ferromagnetic atoms.

The ferromagnetic particles in a polymer matrix can interact either via magnetic dipole-dipole interactions, which are comparatively long range, or by the exchange forces which come into play if the particles are in direct contact with each other and electron wave functions overlap. If these interacting forces are strong, the whole composite will be ferromagnetic, i.e. the moments of neighboring nanoparticles will be coupled providing a long-range magnetic order. In the opposite case the moments of neighboring particles interact weakly, the magnetic order breaks down and the magnetic momentum of each nanoparticle fluctuates independently. This latter state is known as superparamagnetic, analogous to the paramagnetic state of atoms in solid bodies. The other parameter which determines the magnetic state of a composite is temperature, which works against the magnetic ordering and leads to the presence of a threshold between the ferromagnetic and the superparamagnetic state at a particular temperature. The parameter characterising interparticle interaction is the magnetic moment relaxation time τ_r which can be determined by a number of techniques [46].

The question of whether a system is superparamagnetic or ferromagnetic is crucial for the problem of magnetic media storage density, since if information is to be stored in each single-domain particle, its momentum must be uncoupled from the neighboring ones, i.e. the system as a whole must be superparamagnetic. Another important question is how small the anisotropic nanoparticle can be to allow reliable storage of information (momentum up or down in the case of easy magnetization axe existence). There are a number of techniques which can be applied to approach the outlined problems.

Measurement of the magnetization (measurement of magnetic moment of the sample dependence on the magnetic field strength) is the most straightforward and simple to understand. It gives information about the magnetic moments statics when the measurements are conducted in a constant magnetic field or about the low frequency dynamics if pulsed magnetic fields are used.

Mössbauer spectroscopy is an alternative and very sensitive method, since all the nuclei transitions are very sensitive to the local magnetic fields which affect the dipole and quadrupole momenta of the nucleus. However, Mössbauer spectroscopy gives information about the magnetic momentum dynamics, since what is actually obtained is the magnetic momentum averaged along the Mössbauer measurement time τ_m ($\tau_m \sim 10^{-8}$ s for ^{57}Fe in particles with uniaxial anisotropy [47]). Thus Mössbauer spectroscopy only provides information about the magnetic momenta relaxation dynamics within the time scale of 10^{-8} s.

Magnetic force microscopy (MFM) is a recently developed convenient means to study the local fields by using an AFM tip made of ferromagnetic material in the tapping mode. The resolution of MFM is of the order of units of nanometers, which allows mapping of the local magnetic fields of the nanoparticles in a polymer matrix.

If we examine the Mössbauer spectra (Fig. 25) of the PVA-Fe$_3$O$_4$ nanocomposites described above with 150 wt % concentration of 20-nm filler particles, we can see that following a decrease in temperature from 290 K (A) to 200 K (B) and then to 77 K (C) the badly resolved doublet, typical for the superparamagnetic state, is substituted by a well-resolved sextet which is evidence for the ferromagnetic state of the system. Such behavior is typical for the superparamagnetic-ferromagnetic transition which takes place in this case at a temperature of approx. 200 K, at which the intensity of the doublet and the sextet are of the same order.

If one examines the corresponding magnetization dependencies on the magnetic field (Fig. 26) for nanocomposites with 120 and 150 wt % (correspondingly 27 and 35 vol %) of Fe$_3$O$_4$, it can be seen that at room temperature both are in the superparamagnetic state. The absence of hysteresis on the magnetization curve is straightforward evidence of this fact. Thus both the

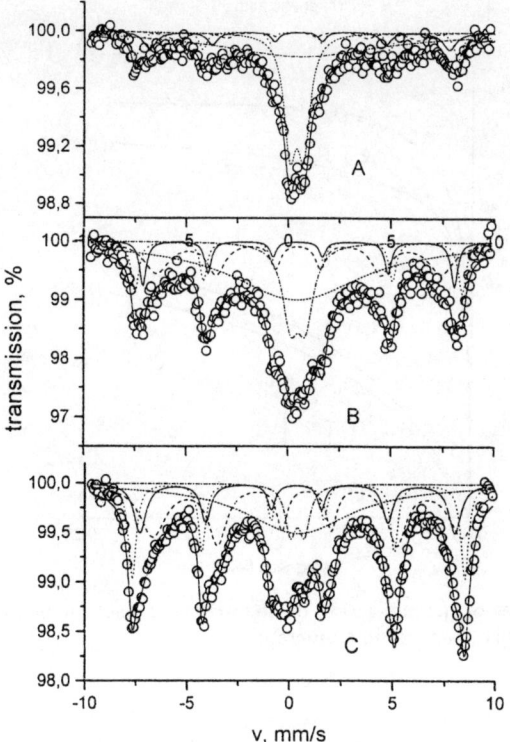

Fig. 25. Mössbauer spectra of PVA-Fe$_3$O$_4$ nanocomposites, 150 wt % concentration of 20-nm size filler particles, at *A* 290 K, *B* 200 K and *C* 77 K

dynamic and static measurements of magnetic momenta interaction lead to the conclusion that the system is in the superparamagnetic state at room temperature.

One of the consequences of this fact is that each particle can be used as a 1-bit information carrier, if the reliability of storage is good enough (see above). This is quite an amazing perspective indeed, since the information storage density, if it can be stored by each nanoparticle, equals ca. 10^{17} bit/cm^2=10,000 Gbit/cm^2, which is a very impressive number indeed. Nevertheless, there are still many obstacles in the way of nanocomposite magnetic storage media manufacture. First of all the information must be stored reliably by each nanoparticle and secondly the particles must form relatively well-ordered 2-D arrays in the matrix. The first problem can be solved by increasing the particles anisotropy, and the second by choosing appropriate synthetic conditions [48]. The possibility of manufacturing MRAM based on composites of ferromagnetic nanoparticles in conductive matrices is another striking possibility to be explored in the near future.

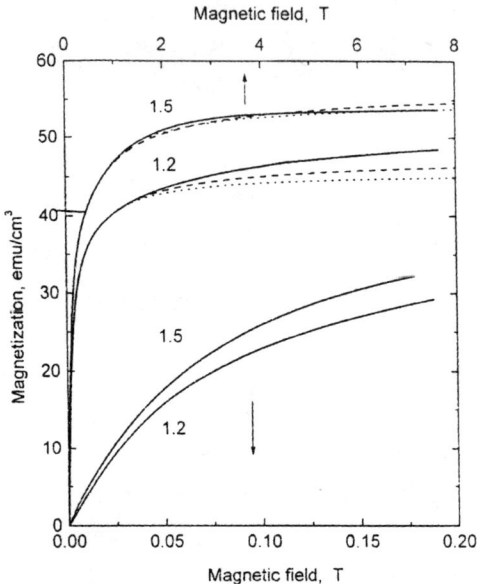

Fig. 26. Dependencies of magnetization on the magnetic field for nanocomposites with 120 and 150 wt % of Fe_3O_4 at room temperature

8
Conclusions

Summarizing the reviewed facts and trends in the development of devices based on polymer-nanocomposites, the opinion can be expressed that, in contrast to the microelectronic circuitry, operating based on the ordered functional elements such as logical gates, in the case of polymer-nanocomposites we can utilize the properties of disordered systems, since the nanoparticles are stochastically dispersed in the matrix. Above we described how the properties of disordered systems, such as the presence of the percolation threshold, change in structure and the dependence of the topology of the interconnected chains on the filler concentration, ability to predictably synthesize the infinite clusters with the programmable topology of the conductive network, can be utilized for LEDs, photodiodes, solar cells and the manufacture of gas sensors. It is quite remarkable, therefore, that the properties of disordered media can be utilized for device applications.

The second conclusion to be made is that devices that utilize the properties of low dimensional objects such as nanoparticles – quantum dots – are promising due to the possibility of tailoring a number of electrophysical, optical and magnetic properties changing the size of the nanoparticles, which can be controlled during the synthesis. Modification of the nanoparticle surface, the possibility of doping and the opportunity to fill the matrix with nanoparticles of both the p- and n-type also broaden the perspective of the device applications of polymer-nanocomposites.

9
Future Perspectives of Nanocomposite-Based Devices

In the previous sections recent developments in the field of polymer-nanocomposite devices have been reviewed. But what are the perspectives of the application of polymer-nanocomposite devices in the near future? Let us now be somewhat futuristic in our prognosis.

One direction which seems to be promising is based on the possibility of building ordered arrays of nanoparticles in the polymer matrix [49]. The simultaneous evaporation of sulfides along with the chemical vapor deposition of polymers (Sect. 2.7) gives under certain conditions ordered two-dimensional arrays of nanoparticles. If a narrow distribution of particle sizes can be reached, a number of opportunities arise for the manufacture of electronic circuits.

A number of possibilities exists to manufacture the nanocomposite circuit boards. The first possibility is to convert non-conductive poly(paraxylylene), obtained as described in Sect. 2.7, into semiconducting poly(paraphenylenevinylene) by heating with an electron beam. The second possibility consists of the light-induced doping of polyaniline-based nanocomposites, allowing a change in the conductivity of the polyaniline from an insulating phase into a semiconductor or even metallic depending on the extent of doping.

Using these two methods nanoscale circuits, consisting of conductors, semiconductors and isolating areas, can be obtained. Elements such as tunnel resonant diodes and single electron transistors can be the structural elements of such circuits, analogous to devices described in the literature [50].

A report of a single electron transistor made operable based on only one CdSe nanoparticle junction between two gold current leads produced by e-beam lithography [53] demonstrates the possibility of making a single-nanoparticle analogue of a silicon-based device, such as a tunnel resonant diode, using colloid chemistry as a means for particles formation instead of microelectronics technology.

Another opportunity consists of the preparation of ordered arrays of nanoparticles by templating them to DNA molecules [51], bacterial self-assembled layers [52] or micelle-like structures [53] in order to form two- or three-dimensional ordered systems. Using such biologically based templates, it is possible to assemble quite complex arrays of functional nanoparticles for electronic and optolelectronic applications.

An even more attractive method exists to utilize polymer-nanocomposites for neural networks applications. Nanoparticles can be used as knots in a network, and connections between these knots can be established by local electrochemical doping, which converts the polymer from an insulating into a conductive state. Since electrochemical doping is a reversible process one can either establish or break connections between knots, allowing the learning function and adaptivity of the networks.

Anther promising area of development is optoelectronics and optical computing. Nanocomposite materials allow so-called sub-λ light sources to be assembled, i.e. light emitters with a size smaller than the emitted light wavelength. In this way, optoelectronic high integration circuits with nanoscale LEDs and photodiodes can be manufactured.

The single-domain nature and superparamagnetic behavior of nanoparticles containing ferromagnetic metals could possibly be utilized for optical computing and for magneto-optical storage media manufacturing, utilizing the Faraday effect, i.e. the change of light polarization in a magnetic field in order to read the information. Changing the magnetization of the superparamagnetic nanoparticles in the desired way can result in a modulation of the light transmitted through such media with the spatial resolution restricted only by the light wavelength. The development of spin electronics and devices such as MRAM is another prospective area for the application of polymer-nanocomposites containing ordered arrays of ferromagnetic nanoparticles.

References

1. http://www.cdtltd.co.uk/technology/
2. Su WP, Schreiffer JR, Heeger AJ (1980) Phys Rev B 224:2099
3. Sariciftci S (ed) (1998) Semiconductor band versus exciton. Pergamon
4. Roman LS, Mammo W, Pettersson LA, Andersson MR, Inganäs O (1998) Adv Mater 10:774
5. http://www.cdtltd.co.uk/technology/seikotvtech.htm
6. International Patent no (ICMR Co. Japan)
7. International Patent no (ICMR Co., Japan)
8. Godovsky D, Zaretsky D, Kundig A, Caseri W (1998) Proc ECOS 98, p 26
9. Tian Y, Newton T, Kotov N, Guldi D, Fendler J (1996) J Phys Chem 100:8927
10. Trakhtenberg L, Gerasimov G, Grigor'ev E (1999) Russ J Phys Chem 73:209
11. Pomogailo AD (1997) Usp Khim 66:750 (in russian)
12. Gardenas TG, Munoz DC (1993) Macromol Chem 194:3377
13. Olsen AW, Kafafi ZH (1993) J Am Chem Soc 113:7758
14. Heilmann A, Kampfrath G, Hopfs V (1988) J Phys D 21:986
15. Gerasimov G, Sochilin V, Chvalun S, Volkova L, Kardash I (1996) Macromol Chem Phys 197:1387
16. Alexandrova L, Sochilin V, Gerasimov G, Kardash I (1997) Polymer 38:271
17. Brus L (1986) J Phys Chem 90:2555
18. Wang Y, Herron N (1991) J Phys Chem 95:525
19. Godovsky DY (1995) App Polym Sci 119:81
20. Varfolomeev AE, Godovsky DY, Zaretsky DF, Volkov A, Moskvina M (1995) JETP Lett 62:344
21. Godovsky D, Sukharev V, Volkov A, Moskvina M (1993) Phys Chem Solids 54:1613
22. Shklovski BI, Efros AL (1979) Physics of doped semiconductors. Nauka, Moscow (in russian)
23. Yu G, Gao J, Hummelen JC, Wudl F, Heeger AJ (1995) Science 270:1789
24. Godovsky D unpublished results
25. Yu G, Heeger AJ (1995) J Appl Phys 78:4510

26. Godovsky D, Inganäs O Appl Phys Lett, to be published
27. Yu G, Gao J, Hummelen JC, Wudl F, Heeger AJ (1995) Science 270:1789
28. Friend RH, Denton GJ, Halls JJM (1997) Synth Metals 84:463
29. Burroughes JH, Forest SR (1990) Nature 347:539
30. Granström M, Berggren M, Inganäs O (1995) Science 267 :1479
31. Colvin VL, Schlamp MC, Alivisatos AP (1994) Nature 370:354
32. Berggren M, Gustafsson G, Inganäs O, Andersson MR, Wennerström O, Hjertberg T (1994) Nature 372:444
33. Schlamp MC, Peng X, Alivisatos AP (1997) J Appl Phys 82:5837
34. Pede D, Smela E, Johansson T, Johansson M, Inganäs O (1998) Adv Mat 10:233
35. Bulovic V, Burrows PE, Garbuzov DZ, Forrest SR (1997) In: McConnell RD (ed) Future generation photovoltaic technologies. AIP Press, Woodbury, NY, pp 235, 404
36. Sariciftci NS, Braun D, Zhang C, Srdanov VI, Heeger AJ, Stucky G, Wudl F.(1993) Appl Phys Lett 62:585
37. Godovsky D, Varfolomeev A, Zaretsky D, Chandrakhati N, Kundig A, Caseri W, Smith P (1999) Adv Mat, in press
38. Greenham NC, Peng XG, Alivisatos AP (1996) Phys Rev B 54:17628
39. Godovsky D, Varfolomeev A, Zaretsky D, Kundig A, Caseri W (1998) ECOS 98, Cadarache, France, Abstracts, p 21
40. Godovsky D, Volkov A, Sukharev V, Moskvina M (1994) Analyst 118:997
41. Godovsky D (1993) PhD thesis
42. Trakhtenberg L, Gerasimov G, Grigoriev E (1996) In: Durig J, Klabunde K (eds) 2nd Intl Conf on Low Temperature Chemistry Book Mark Press, p 221
43. Sergeev G, Zagorsky V, Petrukhin M, Zavialov S, Grigor'ev E, Trakhtenberg L (1997) Anal Commun 34:113
44. Gerasimov G, Grigor'ev E, Grigoriev A (1998) Chim Fiz 17:180
45. Godovsky D, Varfolomeev A, Kapystin G, Cherepanov V, Efremova D (1999) Adv Mat Opt Electron, in print
46. Morup S (1983) J. Magn Magn Mater 37:39
47. Morup S, Tronc E (1994) Phys Rev Lett 72 :3278
48. Caseri W private communication
49. Mirkin CA, Letsinger RL, Mucic R, Storhoff JJ (1996) Nature 382:607
50. Klein D, Roth R, Lim A, Alivisatos AP, McEuen PL (1997) Nature 389:699
51. Spatz JP, Roesher A, Möller M (1996) Adv Mater 8:334
52. Shenton W, Pum D, Sleytr U, Mann S (1997) Nature 389:585
53. Klein D, Roth R, Lim A, Alivisatos P, Mceuen P (1997) Nature 389:609

Editor: Prof. U.W. Suter
Received: September 1999

Author Index Volumes 101–153

Author Index Volumes 1–100 see Volume 100

de, Abajo, J. and de la Campa, J.G.: Processable Aromatic Polyimides. Vol. 140, pp. 23-60.
Adolf, D. B. see Ediger, M. D.: Vol. 116, pp. 73-110.
Aharoni, S. M. and *Edwards, S. F.*: Rigid Polymer Networks. Vol. 118, pp. 1-231.
Améduri, B., Boutevin, B. and *Gramain, P.*: Synthesis of Block Copolymers by Radical Polymerization and Telomerization. Vol. 127, pp. 87-142.
Améduri, B. and *Boutevin, B.*: Synthesis and Properties of Fluorinated Telechelic Monodispersed Compounds. Vol. 102, pp. 133-170.
Amselem, S. see Domb, A. J.: Vol. 107, pp. 93-142.
Andrady, A. L.: Wavelenght Sensitivity in Polymer Photodegradation. Vol. 128, pp. 47-94.
Andreis, M. and *Koenig, J. L.*: Application of Nitrogen-15 NMR to Polymers. Vol. 124, pp. 191-238.
Angiolini, L. see Carlini, C.: Vol. 123, pp. 127-214.
Anseth, K. S., Newman, S. M. and *Bowman, C. N.*: Polymeric Dental Composites: Properties and Reaction Behavior of Multimethacrylate Dental Restorations. Vol. 122, pp. 177-218.
Antonietti, M. see Cölfen, H.: Vol. 150, pp. 67-187.
Armitage, B. A. see O'Brien, D. F.: Vol. 126, pp. 53-58.
Arndt, M. see Kaminski, W.: Vol. 127, pp. 143-187.
Arnold Jr., F. E. and *Arnold, F. E.*: Rigid-Rod Polymers and Molecular Composites. Vol. 117, pp. 257-296.
Arshady, R.: Polymer Synthesis via Activated Esters: A New Dimension of Creativity in Macromolecular Chemistry. Vol. 111, pp. 1-42.

Bahar, I., Erman, B. and *Monnerie, L.*: Effect of Molecular Structure on Local Chain Dynamics: Analytical Approaches and Computational Methods. Vol. 116, pp. 145-206.
Ballauff, M. see Dingenouts, N.: Vol. 144, pp. 1-48.
Baltá-Calleja, F. J., González Arche, A., Ezquerra, T. A., Santa Cruz, C., Batallón, F., Frick, B. and *López Cabarcos, E.*: Structure and Properties of Ferroelectric Copolymers of Poly(vinylidene) Fluoride. Vol. 108, pp. 1-48.
Barshtein, G. R. and *Sabsai, O. Y.*: Compositions with Mineralorganic Fillers. Vol. 101, pp.1-28.
Baschnagel, J., Binder, K., Doruker, P., Gusev, A. A., Hahn, O., Kremer, K., Mattice, W. L., Müller-Plathe, F., Murat, M., Paul, W., Santos, S., Sutter, U. W., Tries, V.: Bridging the Gap Between Atomistic and Coarse-Grained Models of Polymers: Status and Perspectives. Vol. 152, pp. 41-156.
Batallán, F. see Baltá-Calleja, F. J.: Vol. 108, pp. 1-48.
Batog, A. E., Pet'ko, I. P., Penczek, P.: Aliphatic-Cycloaliphatic Epoxy Compounds and Polymers. Vol. 144, pp. 49-114.
Barton, J. see Hunkeler, D.: Vol. 112, pp. 115-134.
Bell, C. L. and *Peppas, N. A.*: Biomedical Membranes from Hydrogels and Interpolymer Complexes. Vol. 122, pp. 125-176.
Bellon-Maurel, A. see Calmon-Decriaud, A.: Vol. 135, pp. 207-226.
Bennett, D. E. see O'Brien, D. F.: Vol. 126, pp. 53-84.

Berry, G.C.: Static and Dynamic Light Scattering on Moderately Concentraded Solutions: Isotropic Solutions of Flexible and Rodlike Chains and Nematic Solutions of Rodlike Chains. Vol. 114, pp. 233-290.
Bershtein, V. A. and *Ryzhov, V. A.*: Far Infrared Spectroscopy of Polymers. Vol. 114, pp. 43-122.
Bigg, D. M.: Thermal Conductivity of Heterophase Polymer Compositions. Vol. 119, pp. 1-30.
Binder, K.: Phase Transitions in Polymer Blends and Block Copolymer Melts: Some Recent Developments. Vol. 112, pp. 115-134.
Binder, K.: Phase Transitions of Polymer Blends and Block Copolymer Melts in Thin Films. Vol. 138, pp. 1-90.
Binder, K. see Baschnagel, J.: Vol. 152, pp. 41-156.
Bird, R. B. see Curtiss, C. F.: Vol. 125, pp. 1-102.
Biswas, M. and *Mukherjee, A.*: Synthesis and Evaluation of Metal-Containing Polymers. Vol. 115, pp. 89-124.
Bolze, J. see Dingenouts, N.: Vol. 144, pp. 1-48.
Boutevin, B. and *Robin, J. J.*: Synthesis and Properties of Fluorinated Diols. Vol. 102. pp. 105-132.
Boutevin, B. see Amédouri, B.: Vol. 102, pp. 133-170.
Boutevin, B. see Améduri, B.: Vol. 127, pp. 87-142.
Bowman, C. N. see Anseth, K. S.: Vol. 122, pp. 177-218.
Boyd, R. H.: Prediction of Polymer Crystal Structures and Properties. Vol. 116, pp. 1-26.
Briber, R. M. see Hedrick, J. L.: Vol. 141, pp. 1-44.
Bronnikov, S. V., Vettegren, V. I. and *Frenkel, S. Y.*: Kinetics of Deformation and Relaxation in Highly Oriented Polymers. Vol. 125, pp. 103-146.
Bruza, K. J. see Kirchhoff, R. A.: Vol. 117, pp. 1-66.
Budkowski, A.: Interfacial Phenomena in Thin Polymer Films: Phase Coexistence and Segregation. Vol. 148, pp. 1-112.
Burban, J. H. see Cussler, E. L.: Vol. 110, pp. 67-80.
Burchard, W.: Solution Properties of Branched Macromolecules. Vol. 143, pp. 113-194.

Calmon-Decriaud, A. Bellon-Maurel, V., Silvestre, F.: Standard Methods for Testing the Aerobic Biodegradation of Polymeric Materials. Vol 135, pp. 207-226.
Cameron, N. R. and *Sherrington, D. C.*: High Internal Phase Emulsions (HIPEs)-Structure, Properties and Use in Polymer Preparation. Vol. 126, pp. 163-214.
de la Campa, J. G. see de Abajo, , J.: Vol. 140, pp. 23-60.
Candau, F. see Hunkeler, D.: Vol. 112, pp. 115-134.
Canelas, D. A. and *DeSimone, J. M.*: Polymerizations in Liquid and Supercritical Carbon Dioxide. Vol. 133, pp. 103-140.
Capek, I.: Kinetics of the Free-Radical Emulsion Polymerization of Vinyl Chloride. Vol. 120, pp. 135-206.
Capek, I.: Radical Polymerization of Polyoxyethylene Macromonomers in Disperse Systems. Vol. 145, pp. 1-56.
Capek, I.: Radical Polymerization of Polyoxyethylene Macromonomers in Disperse Systems. Vol. 146, pp. 1-56.
Carlini, C. and *Angiolini, L.*: Polymers as Free Radical Photoinitiators. Vol. 123, pp. 127-214.
Carter, K. R. see Hedrick, J. L.: Vol. 141, pp. 1-44.
Casas-Vazquez, J. see Jou, D.: Vol. 120, pp. 207-266.
Chandrasekhar, V.: Polymer Solid Electrolytes: Synthesis and Structure. Vol 135, pp. 139-206
Chang, J. Y. see Han, M. J.: Vol. 153, pp. 1-36
Charleux, B., Faust R.: Synthesis of Branched Polymers by Cationic Polymerization. Vol. 142, pp. 1-70.
Chen, P. see Jaffe, M.: Vol. 117, pp. 297-328.
Choe, E.-W. see Jaffe, M.: Vol. 117, pp. 297-328.
Chow, T. S.: Glassy State Relaxation and Deformation in Polymers. Vol. 103, pp. 149-190.
Chung, T.-S. see Jaffe, M.: Vol. 117, pp. 297-328.

Cölfen, H. and *Antonietti, M.*: Field-Flow Fractionation Techniques for Polymer and Colloid Analysis. Vol. 150, pp. 67-187.
Comanita, B. see Roovers, J.: Vol. 142, pp. 179-228.
Connell, J. W. see Hergenrother, P. M.: Vol. 117, pp. 67-110.
Criado-Sancho, M. see Jou, D.: Vol. 120, pp. 207-266.
Curro, J.G. see Schweizer, K.S.: Vol. 116, pp. 319-378.
Curtiss, C. F. and *Bird, R. B.*: Statistical Mechanics of Transport Phenomena: Polymeric Liquid Mixtures. Vol. 125, pp. 1-102.
Cussler, E. L., Wang, K. L. and *Burban, J. H.*: Hydrogels as Separation Agents. Vol. 110, pp. 67-80.

DeSimone, J. M. see Canelas D. A.: Vol. 133, pp. 103-140.
DiMari, S. see Prokop, A.: Vol. 136, pp. 1-52.
Dimonie, M. V. see Hunkeler, D.: Vol. 112, pp. 115-134.
Dingenouts, N., Bolze, J., Pötschke, D., Ballauf, M.: Analysis of Polymer Latexes by Small-Angle X-Ray Scattering. Vol. 144, pp. 1-48
Dodd, L. R. and *Theodorou, D. N.*: Atomistic Monte Carlo Simulation and Continuum Mean Field Theory of the Structure and Equation of State Properties of Alkane and Polymer Melts. Vol. 116, pp. 249-282.
Doelker, E.: Cellulose Derivatives. Vol. 107, pp. 199-266.
Dolden, J. G.: Calculation of a Mesogenic Index with Emphasis Upon LC-Polyimides. Vol. 141, pp. 189-245.
Domb, A. J., Amselem, S., Shah, J. and *Maniar, M.*: Polyanhydrides: Synthesis and Characterization. Vol.107, pp. 93-142.
Doruker, P. see Baschnagel, J.: Vol. 152, pp. 41-156.
Dubois, P. see Mecerreyes, D.: Vol. 147, pp. 1-60.
Dubrovskii, S. A. see Kazanskii, K. S.: Vol. 104, pp. 97-134.
Dunkin, I. R. see Steinke, J.: Vol. 123, pp. 81-126.
Dunson, D. L. see McGrath, J. E.: Vol. 140, pp. 61-106.

Eastmond, G. C.: Poly(ε-caprolactone) Blends. Vol.149, pp. 59-223.
Economy, J. and *Goranov, K.*: Thermotropic Liquid Crystalline Polymers for High Performance Applications. Vol. 117, pp. 221-256.
Ediger, M. D. and *Adolf, D. B.*: Brownian Dynamics Simulations of Local Polymer Dynamics. Vol. 116, pp. 73-110.
Edwards, S. F. see Aharoni, S. M.: Vol. 118, pp. 1-231.
Endo, T. see Yagci, Y.: Vol. 127, pp. 59-86.
Engelhardt, H. and *Grosche, O.*: Capillary Electrophoresis in Polymer Analysis. Vol. 150, pp. 189-217.
Erman, B. see Bahar, I.: Vol. 116, pp. 145-206.
Ewen, B, Richter, D.: Neutron Spin Echo Investigations on the Segmental Dynamics of Polymers in Melts, Networks and Solutions. Vol. 134, pp. 1-130.
Ezquerra, T. A. see Baltá-Calleja, F. J.: Vol. 108, pp. 1-48.

Faust, R. see Charleux, B: Vol. 142, pp. 1-70.
Fekete, E see Pukánszky, B: Vol. 139, pp. 109-154.
Fendler, J.H.: Membrane-Mimetic Approach to Advanced Materials. Vol. 113, pp. 1-209.
Fetters, L. J. see Xu, Z.: Vol. 120, pp. 1-50.
Förster, S. and *Schmidt, M.*: Polyelectrolytes in Solution. Vol. 120, pp. 51-134.
Freire,J.J.: Conformational Properties of Branched Polymers: Theory and Simulations. Vol. 143, pp. 35-112.
Frenkel, S. Y. see Bronnikov, S. V.: Vol. 125, pp. 103-146.
Frick, B. see Baltá-Calleja, F. J.: Vol. 108, pp. 1-48.
Fridman, M. L.: see Terent´eva, J. P.: Vol. 101, pp. 29-64.
Funke, W.: Microgels-Intramolecularly Crosslinked Macromolecules with a Globular Structure. Vol. 136, pp. 137-232.

Galina, H.: Mean-Field Kinetic Modeling of Polymerization: The Smoluchowski Coagulation Equation. Vol. 137, pp. 135-172.
Ganesh, K. see *Kishore, K.:* Vol. 121, pp. 81-122.
Gaw, K. O. and *Kakimoto, M.:* Polyimide-Epoxy Composites. Vol. 140, pp. 107-136.
Geckeler, K. E. see *Rivas, B.:* Vol. 102, pp. 171-188.
Geckeler, K. E.: Soluble Polymer Supports for Liquid-Phase Synthesis. Vol. 121, pp. 31-80.
Gehrke, S. H.: Synthesis, Equilibrium Swelling, Kinetics Permeability and Applications of Environmentally Responsive Gels. Vol. 110, pp. 81-144.
de Gennes, P.-G.: Flexible Polymers in Nanopores. Vol. 138, pp. 91-106.
Giannelis, E.P., Krishnamoorti, R., Manias, E.: Polymer-Silicate Nanocomposites: Model Systems for Confined Polymers and Polymer Brushes. Vol. 138, pp. 107-148.
Godovsky, D. Y.: Device Applications of Polymer-Nanocomposites. Vol. 153, pp. 163-205
Godovsky, D. Y.: Electron Behavior and Magnetic Properties Polymer-Nanocomposites. Vol. 119, pp. 79-122.
González Arche, A. see *Baltá-Calleja, F. J.:* Vol. 108, pp. 1-48.
Goranov, K. see *Economy, J.:* Vol. 117, pp. 221-256.
Gramain, P. see *Améduri, B.:* Vol. 127, pp. 87-142.
Grest, G.S.: Normal and Shear Forces Between Polymer Brushes. Vol. 138, pp. 149-184
Grigorescu, G, Kulicke, W.-M.: Prediction of Viscoelastic Properties and Shear Stability of Polymers in Solution. Vol. 152, p. 1-40.
Grosberg, A. and *Nechaev, S.:* Polymer Topology. Vol. 106, pp. 1-30.
Grosche, O. see *Engelhardt, H.:* Vol. 150, pp. 189-217.
Grubbs, R., Risse, W. and *Novac, B.:* The Development of Well-defined Catalysts for Ring-Opening Olefin Metathesis. Vol. 102, pp. 47-72.
van Gunsteren, W. F. see *Gusev, A. A.:* Vol. 116, pp. 207-248.
Gusev, A. A., Müller-Plathe, F., van Gunsteren, W. F. and *Suter, U. W.:* Dynamics of Small Molecules in Bulk Polymers. Vol. 116, pp. 207-248.
Gusev, A. A. see *Baschnagel, J.:* Vol. 152, pp. 41-156.
Guillot, J. see *Hunkeler, D.:* Vol. 112, pp. 115-134.
Guyot, A. and *Tauer, K.:* Reactive Surfactants in Emulsion Polymerization. Vol. 111, pp. 43-66.

Hadjichristidis, N., Pispas, S., Pitsikalis, M., Iatrou, H., Vlahos, C.: Asymmetric Star Polymers Synthesis and Properties. Vol. 142, pp. 71-128.
Hadjichristidis, N. see *Xu, Z.:* Vol. 120, pp. 1-50.
Hadjichristidis, N. see *Pitsikalis, M.:* Vol. 135, pp. 1-138.
Hahn, O. see *Baschnagel, J.:* Vol. 152, pp. 41-156.
Hall, H. K. see *Penelle, J.:* Vol. 102, pp. 73-104.
Höcker, H. see *Klee, D.:* Vol. 149, pp. 1-57.
Hammouda, B.: SANS from Homogeneous Polymer Mixtures: A Unified Overview. Vol. 106, pp. 87-134.
Han, M.J. and *Chang, J.Y.:* Polynucleotide Analogues. Vol. 153, pp. 1-36
Harada, A.: Design and Construction of Supramolecular Architectures Consisting of Cyclodextrins and Polymers. Vol. 133, pp. 141-192.
Haralson, M. A. see *Prokop, A.:* Vol. 136, pp. 1-52.
Hassan, C.M. and *Peppas, N.A.:* Structure and Applications of Poly(vinyl alcohol) Hydrogels Produced by Conventional Crosslinking or by Freezing/Thawing Methods. Vol. 153, pp. 37-65
Hawker, C. J. Dentritic and Hyperbranched Macromolecules – Precisely Controlled Macromolecular Architectures. Vol. 147, pp. 113-160
Hawker, C. J. see *Hedrick, J. L.:* Vol. 141, pp. 1-44.
Hedrick, J. L., Carter, K. R., Labadie, J. W., Miller, R. D., Volksen, W., Hawker, C. J., Yoon, D. Y., Russell, T. P., McGrath, J. E., Briber, R. M.: Nanoporous Polyimides. Vol. 141, pp. 1-44.
Hedrick, J. L., Labadie, J. W., Volksen, W. and *Hilborn, J. G.:* Nanoscopically Engineered Polyimides. Vol. 147, pp. 61-112.

Hedrick, J. L. see *Hergenrother, P. M.*: Vol. 117, pp. 67-110.
Hedrick, J. L. see *Kiefer, J.*: Vol. 147, pp. 161-247.
Hedrick, J.L. see *McGrath, J. E.*: Vol. 140, pp. 61-106.
Heller, J.: Poly (Ortho Esters). Vol. 107, pp. 41-92.
Hemielec, A. A. see *Hunkeler, D.*: Vol. 112, pp. 115-134.
Hergenrother, P. M., Connell, J. W., Labadie, J. W. and *Hedrick, J. L.*: Poly(arylene ether)s Containing Heterocyclic Units. Vol. 117, pp. 67-110.
Hernández-Barajas, J. see *Wandrey, C.*: Vol. 145, pp. 123-182.
Hervet, H. see *Léger, L.*: Vol. 138, pp. 185-226.
Hilborn, J. G. see *Hedrick, J. L.*: Vol. 147, pp. 61-112.
Hilborn, J. G. see *Kiefer, J.*: Vol. 147, pp. 161-247.
Hiramatsu, N. see *Matsushige, M.*: Vol. 125, pp. 147-186.
Hirasa, O. see *Suzuki, M.*: Vol. 110, pp. 241-262.
Hirotsu, S.: Coexistence of Phases and the Nature of First-Order Transition in Poly-N-isopropylacrylamide Gels. Vol. 110, pp. 1-26.
Hamley, I. W.: Crystallization in Block Copolymers. Vol. 148, pp. 113-138.
Hornsby, P.: Rheology, Compoundind and Processing of Filled Thermoplastics. Vol. 139, pp. 155-216.
Hult, A., Johansson, M., Malmström, E.: Hyperbranched Polymers. Vol. 143, pp. 1-34.
Hunkeler, D., Candau, F., Pichot, C., Hemielec, A. E., Xie, T. Y., Barton, J., Vaskova, V., Guillot, J., Dimonie, M. V., Reichert, K. H.: Heterophase Polymerization: A Physical and Kinetic Comparision and Categorization. Vol. 112, pp. 115-134.
Hunkeler, D. see *Prokop, A.*: Vol. 136, pp. 1-52; 53-74.
Hunkeler, D see *Wandrey, C.*: Vol. 145, pp. 123-182.

Iatrou, H. see *Hadjichristidis, N.*: Vol. 142, pp. 71-128
Ichikawa, T. see *Yoshida, H.*: Vol. 105, pp. 3-36.
Ihara, E. see *Yasuda, H.*: Vol. 133, pp. 53-102.
Ikada, Y. see *Uyama, Y.*: Vol. 137, pp. 1-40.
Ilavsky, M.: Effect on Phase Transition on Swelling and Mechanical Behavior of Synthetic Hydrogels. Vol. 109, pp. 173-206.
Imai, Y.: Rapid Synthesis of Polyimides from Nylon-Salt Monomers. Vol. 140, pp. 1-23.
Inomata, H. see *Saito, S.*: Vol. 106, pp. 207-232.
Inoue, S. see *Sugimoto, H.*: Vol. 146, pp. 39-120.
Irie, M.: Stimuli-Responsive Poly(N-isopropylacrylamide), Photo- and Chemical-Induced Phase Transitions. Vol. 110, pp. 49-66.
Ise, N. see *Matsuoka, H.*: Vol. 114, pp. 187-232.
Ito, K., Kawaguchi, S,:Poly(macronomers), Homo- and Copolymerization. Vol. 142, pp. 129-178.
Ivanov, A. E. see *Zubov, V. P.*: Vol. 104, pp. 135-176.

Jacob, S. and *Kennedy, J.*: Synthesis, Characterization and Properties of OCTA-ARM Polyisobutylene-Based Star Polymers. Vol. 146, pp. 1-38.
Jaffe, M., Chen, P., Choe, E.-W., Chung, T.-S. and *Makhija, S.*: High Performance Polymer Blends. Vol. 117, pp. 297-328.
Jancar, J.: Structure-Property Relationships in Thermoplastic Matrices. Vol. 139, pp. 1-66.
Jerôme, R.: see *Mecerreyes, D.*: Vol. 147, pp. 1-60.
Jiang, M., Li, M., Xiang, M. and *Zhou, H.*: Interpolymer Complexation and Miscibility and Enhancement by Hydrogen Bonding. Vol. 146, pp. 121-194.
Johansson, M. see *Hult, A.*: Vol. 143, pp. 1-34.
Joos-Müller, B. see *Funke, W.*: Vol. 136, pp. 137-232.
Jou, D., Casas-Vazquez, J. and *Criado-Sancho, M.*: Thermodynamics of Polymer Solutions under Flow: Phase Separation and Polymer Degradation. Vol. 120, pp. 207-266.

Kaetsu, I.: Radiation Synthesis of Polymeric Materials for Biomedical and Biochemical Applications. Vol. 105, pp. 81-98.

Kakimoto, M. see Gaw, K. O.: Vol. 140, pp. 107-136.
Kaminski, W. and *Arndt, M.*: Metallocenes for Polymer Catalysis. Vol. 127, pp. 143-187.
Kammer, H. W., Kressler, H. and *Kummerloewe, C.*: Phase Behavior of Polymer Blends - Effects of Thermodynamics and Rheology. Vol. 106, pp. 31-86.
Kandyrin, L. B. and *Kuleznev, V. N.*: The Dependence of Viscosity on the Composition of Concentrated Dispersions and the Free Volume Concept of Disperse Systems. Vol. 103, pp. 103-148.
Kaneko, M. see Ramaraj, R.: Vol. 123, pp. 215-242.
Kang, E. T., Neoh, K. G. and *Tan, K. L.*: X-Ray Photoelectron Spectroscopic Studies of Electroactive Polymers. Vol. 106, pp. 135-190.
Kato, K. see Uyama, Y.: Vol. 137, pp. 1-40.
Kawaguchi, S. see Ito, K.: Vol. 142, p 129-178.
Kazanskii, K. S. and *Dubrovskii, S. A.*: Chemistry and Physics of „Agricultural" Hydrogels. Vol. 104, pp. 97-134.
Kennedy, J. P. see Jacob, S.: Vol. 146, pp. 1-38.
Kennedy, J. P. see Majoros, I.: Vol. 112, pp. 1-113.
Khokhlov, A., Starodybtzev, S. and *Vasilevskaya, V.*: Conformational Transitions of Polymer Gels: Theory and Experiment. Vol. 109, pp. 121-172.
Kiefer, J., Hedrick J. L. and *Hiborn, J. G.*: Macroporous Thermosets by Chemically Induced Phase Separation. Vol. 147, pp. 161-247.
Kilian, H. G. and *Pieper, T.*: Packing of Chain Segments. A Method for Describing X-Ray Patterns of Crystalline, Liquid Crystalline and Non-Crystalline Polymers. Vol. 108, pp. 49-90.
Kim, J. see Quirk, R.P.: Vol. 153, pp. 67-162
Kishore, K. and *Ganesh, K.*: Polymers Containing Disulfide, Tetrasulfide, Diselenide and Ditelluride Linkages in the Main Chain. Vol. 121, pp. 81-122.
Kitamaru, R.: Phase Structure of Polyethylene and Other Crystalline Polymers by Solid-State ^{13}C/MNR. Vol. 137, pp 41-102.
Klee, D. and *Höcker, H.*: Polymers for Biomedical Applications: Improvement of the Interface Compatibility. Vol. 149, pp. 1-57.
Klier, J. see Scranton, A. B.: Vol. 122, pp. 1-54.
Kobayashi, S., Shoda, S. and *Uyama, H.*: Enzymatic Polymerization and Oligomerization. Vol. 121, pp. 1-30.
Köhler, W. and *Schäfer, R.*: Polymer Analysis by Thermal-Diffusion Forced Rayleigh Scattering. Vol. 151, pp. 1-59.
Koenig, J. L. see Andreis, M.: Vol. 124, pp. 191-238.
Koike, T.: Viscoelastic Behavior of Epoxy Resins Before Crosslinking. Vol. 148, pp. 139-188.
Kokufuta, E.: Novel Applications for Stimulus-Sensitive Polymer Gels in the Preparation of Functional Immobilized Biocatalysts. Vol. 110, pp. 157-178.
Konno, M. see Saito, S.: Vol. 109, pp. 207-232.
Kopecek, J. see Putnam, D.: Vol. 122, pp. 55-124.
Koßmehl, G. see Schopf, G.: Vol. 129, pp. 1-145.
Kremer, K. see Baschnagel, J.: Vol. 152, pp. 41-156.
Kressler, J. see Kammer, H. W.: Vol. 106, pp. 31-86.
Kricheldorf, H. R.: Liquid-Cristalline Polyimides. Vol. 141, pp. 83-188.
Krishnamoorti, R. see Giannelis, E.P.: Vol. 138, pp. 107-148.
Kirchhoff, R. A. and *Bruza, K. J.*: Polymers from Benzocyclobutenes. Vol. 117, pp. 1-66.
Kuchanov, S. I.: Modern Aspects of Quantitative Theory of Free-Radical Copolymerization. Vol. 103, pp. 1-102.
Kuchanov, S. I.: Principles of Quantitive Description of Chemical Structure of Synthetic Polymers. Vol. 152, p. 157-202
Kudaibergennow, S.E.: Recent Advances in Studying of Synthetic Polyampholytes in Solutions. Vol. 144, pp. 115-198.
Kuleznev, V. N. see Kandyrin, L. B.: Vol. 103, pp. 103-148.

Kulichkhin, S. G. see Malkin, A. Y.: Vol. 101, pp. 217-258.
Kulicke, W.-M. see Grigorescu, G.: Vol. 152, p. 1-40
Kummerloewe, C. see Kammer, H. W.: Vol. 106, pp. 31-86.
Kuznetsova, N. P. see Samsonov, G. V.: Vol. 104, pp. 1-50.Labadie, J. W. see Hergenrother, P. M.: Vol. 117, pp. 67-110.
Labadie, J. W. see Hedrick, J. L.: Vol. 141, pp. 1-44.
Labadie, J. W. see Hedrick, J. L.: Vol. 147, pp. 61-112.
Lamparski, H. G. see O´Brien, D. F.: Vol. 126, pp. 53-84.
Laschewsky, A.: Molecular Concepts, Self-Organisation and Properties of Polysoaps. Vol. 124, pp. 1-86.
Laso, M. see Leontidis, E.: Vol. 116, pp. 283-318.
Lazár, M. and *RychlΩ, R.*: Oxidation of Hydrocarbon Polymers. Vol. 102, pp. 189-222.
Lechowicz, J. see Galina, H.: Vol. 137, pp. 135-172.
Léger, L., Raphaël, E., Hervet, H.: Surface-Anchored Polymer Chains: Their Role in Adhesion and Friction. Vol. 138, pp. 185-226.
Lenz, R. W.: Biodegradable Polymers. Vol. 107, pp. 1-40.
Leontidis, E., de Pablo, J. J., Laso, M. and *Suter, U. W.*: A Critical Evaluation of Novel Algorithms for the Off-Lattice Monte Carlo Simulation of Condensed Polymer Phases. Vol. 116, pp. 283-318.
Lee, B. see Quirk, R.P: Vol. 153, pp. 67-162
Lee, Y. see Quirk, R.P: Vol. 153, pp. 67-162
Lesec, J. see Viovy, J.-L.: Vol. 114, pp. 1-42.
Li, M. see Jiang, M.: Vol. 146, pp. 121-194.
Liang, G. L. see Sumpter, B. G.: Vol. 116, pp. 27-72.
Lienert, K.-W.: Poly(ester-imide)s for Industrial Use. Vol. 141, pp. 45-82.
Lin, J. and *Sherrington, D. C.*: Recent Developments in the Synthesis, Thermostability and Liquid Crystal Properties of Aromatic Polyamides. Vol. 111, pp. 177-220.
López Cabarcos, E. see Baltá-Calleja, F. J.: Vol. 108, pp. 1-48.

Majoros, I., Nagy, A. and *Kennedy, J. P.*: Conventional and Living Carbocationic Polymerizations United. I. A Comprehensive Model and New Diagnostic Method to Probe the Mechanism of Homopolymerizations. Vol. 112, pp. 1-113.
Makhija, S. see Jaffe, M.: Vol. 117, pp. 297-328.
Malmström, E. see Hult, A.: Vol. 143, pp. 1-34.
Malkin, A. Y. and *Kulichkhin, S. G.*: Rheokinetics of Curing. Vol. 101, pp. 217-258.
Maniar, M. see Domb, A. J.: Vol. 107, pp. 93-142.
Manias, E., see Giannelis, E.P.: Vol. 138, pp. 107-148.
Mashima, K., Nakayama, Y. and *Nakamura, A.*: Recent Trends in Polymerization of a-Olefins Catalyzed by Organometallic Complexes of Early Transition Metals. Vol. 133, pp. 1-52.
Matsumoto, A.: Free-Radical Crosslinking Polymerization and Copolymerization of Multivinyl Compounds. Vol. 123, pp. 41-80.
Matsumoto, A. see Otsu, T.: Vol. 136, pp. 75-138.
Matsuoka, H. and *Ise, N.*: Small-Angle and Ultra-Small Angle Scattering Study of the Ordered Structure in Polyelectrolyte Solutions and Colloidal Dispersions. Vol. 114, pp. 187-232.
Matsushige, K., Hiramatsu, N. and *Okabe, H.*: Ultrasonic Spectroscopy for Polymeric Materials. Vol. 125, pp. 147-186.
Mattice, W. L. see Rehahn, M.: Vol. 131/132, pp. 1-475.
Mattice, W. L. see Baschnagel, J.: Vol. 152, p. 41-156.
Mays, W. see Xu, Z.: Vol. 120, pp. 1-50.
Mays, J.W. see Pitsikalis, M.: Vol.135, pp. 1-138.
McGrath, J. E. see Hedrick, J. L.: Vol. 141, pp. 1-44.
McGrath, J. E., Dunson, D. L., Hedrick, J. L.: Synthesis and Characterization of Segmented Polyimide-Polyorganosiloxane Copolymers. Vol. 140, pp. 61-106.
McLeish, T.C.B., Milner, S. T.: Entangled Dynamics and Melt Flow of Branched Polymers. Vol. 143, pp. 195-256.

Mecerreyes, D., Dubois, P. and *Jerôme, R.*: Novel Macromolecular Architectures Based on Aliphatic Polyesters: Relevance of the „Coordination-Insertion" Ring-Opening Polymerization. Vol. 147, pp. 1 -60.
Mecham, S. J. see *McGrath, J. E.*: Vol. 140, pp. 61-106.
Mikos, A. G. see *Thomson, R. C.*: Vol. 122, pp. 245-274.
Milner, S. T. see *McLeish, T. C. B.*: Vol. 143, pp. 195-256.
Mison, P. and Sillion, B.: Thermosetting Oligomers Containing Maleimides and Nadiimides End-Groups. Vol. 140, pp. 137-180.
Miyasaka, K.: PVA-Iodine Complexes: Formation, Structure and Properties. Vol. 108. pp. 91-130.
Miller, R. D. see *Hedrick, J. L.*: Vol. 141, pp. 1-44.
Monnerie, L. see *Bahar, I.*: Vol. 116, pp. 145-206.
Morishima, Y.: Photoinduced Electron Transfer in Amphiphilic Polyelectrolyte Systems. Vol. 104, pp. 51-96.
Morton M. see *Quirk, R.P*: Vol. 153, pp. 67-162
Mours, M. see *Winter, H. H.*: Vol. 134, pp. 165-234.
Müllen, K. see *Scherf, U.*: Vol. 123, pp. 1-40.
Müller-Plathe, F. see *Gusev, A. A.*: Vol. 116, pp. 207-248.
Müller-Plathe, F. see *Baschnagel, J.*: Vol. 152, p. 41-156.
Mukerherjee, A. see *Biswas, M.*: Vol. 115, pp. 89-124.
Murat, M. see *Baschnagel, J.*: Vol. 152, p. 41-156.
Mylnikov, V.: Photoconducting Polymers. Vol. 115, pp. 1-88.

Nagy, A. see *Majoros, I.*: Vol. 112, pp. 1-11.
Nakamura, A. see *Mashima, K.*: Vol. 133, pp. 1-52.
Nakayama, Y. see *Mashima, K.*: Vol. 133, pp. 1-52.
Narasinham, B., Peppas, N. A.: The Physics of Polymer Dissolution: Modeling Approaches and Experimental Behavior. Vol. 128, pp. 157-208.
Nechaev, S. see *Grosberg, A.*: Vol. 106, pp. 1-30.
Neoh, K. G. see *Kang, E. T.*: Vol. 106, pp. 135-190.
Newman, S. M. see *Anseth, K. S.*: Vol. 122, pp. 177-218.
Nijenhuis, K. te: Thermoreversible Networks. Vol. 130, pp. 1-252.
Noid, D. W. see *Sumpter, B. G.*: Vol. 116, pp. 27-72.
Novac, B. see *Grubbs, R.*: Vol. 102, pp. 47-72.
Novikov, V. V. see *Privalko, V. P.*: Vol. 119, pp. 31-78.

O'Brien, D. F., Armitage, B. A., Bennett, D. E. and *Lamparski, H. G.*: Polymerization and Domain Formation in Lipid Assemblies. Vol. 126, pp. 53-84.
Ogasawara, M.: Application of Pulse Radiolysis to the Study of Polymers and Polymerizations. Vol.105, pp. 37-80.
Okabe, H. see *Matsushige, K.*: Vol. 125, pp. 147-186.
Okada, M.: Ring-Opening Polymerization of Bicyclic and Spiro Compounds. Reactivities and Polymerization Mechanisms. Vol. 102, pp. 1-46.
Okano, T.: Molecular Design of Temperature-Responsive Polymers as Intelligent Materials. Vol. 110, pp. 179-198.
Okay, O. see *Funke, W.*: Vol. 136, pp. 137-232.
Onuki, A.: Theory of Phase Transition in Polymer Gels. Vol. 109, pp. 63-120.
Osad'ko, I.S.: Selective Spectroscopy of Chromophore Doped Polymers and Glasses. Vol. 114, pp. 123-186.
Otsu, T., Matsumoto, A.: Controlled Synthesis of Polymers Using the Iniferter Technique: Developments in Living Radical Polymerization. Vol. 136, pp. 75-138.

de Pablo, J. J. see *Leontidis, E.*: Vol. 116, pp. 283-318.
Padias, A. B. see *Penelle, J.*: Vol. 102, pp. 73-104.
Pascault, J.-P. see *Williams, R. J. J.*: Vol. 128, pp. 95-156.
Pasch, H.: Analysis of Complex Polymers by Interaction Chromatography. Vol. 128, pp. 1-46.

Pasch, H.: Hyphenated Techniques in Liquid Chromatography of Polymers. Vol. 150, pp. 1-66.
Paul, W. see Baschnagel, J.: Vol. 152, p. 41-156.
Penczek, P. see Batog, A. E.: Vol. 144, pp. 49-114.
Penelle, J., Hall, H. K., Padias, A. B. and *Tanaka, H.*: Captodative Olefins in Polymer Chemistry. Vol. 102, pp. 73-104.
Peppas, N. A. see Bell, C. L.: Vol. 122, pp. 125-176.
Peppas, N.A. see Hassan, C.M.: Vol. 153, pp. 37-65
Peppas, N. A. see Narasimhan, B.: Vol. 128, pp. 157-208.
Pet'ko, I. P. see Batog, A. E.: Vol. 144, pp. 49-114.
Pichot, C. see Hunkeler, D.: Vol. 112, pp. 115-134.
Pieper, T. see Kilian, H. G.: Vol. 108, pp. 49-90.
Pispas, S. see Pitsikalis, M.: Vol. 135, pp. 1-138.
Pispas, S. see Hadjichristidis: Vol. 142, pp. 71-128.
Pitsikalis, M., Pispas, S., Mays, J. W., Hadjichristidis, N.: Nonlinear Block Copolymer Architectures. Vol. 135, pp. 1-138.
Pitsikalis, M. see Hadjichristidis: Vol. 142, pp. 71-128.
Pötschke, D. see Dingenouts, N.: Vol 144, pp. 1-48.
Pospíšil, J.: Functionalized Oligomers and Polymers as Stabilizers for Conventional Polymers. Vol. 101, pp. 65-168.
Pospíšil, J.: Aromatic and Heterocyclic Amines in Polymer Stabilization. Vol. 124, pp. 87-190.
Powers, A. C. see Prokop, A.: Vol. 136, pp. 53-74.
Priddy, D. B.: Recent Advances in Styrene Polymerization. Vol. 111, pp. 67-114.
Priddy, D. B.: Thermal Discoloration Chemistry of Styrene-co-Acrylonitrile. Vol. 121, pp. 123-154.
Privalko, V. P. and *Novikov, V. V.*: Model Treatments of the Heat Conductivity of Heterogeneous Polymers. Vol. 119, pp 31-78.
Prokop, A., Hunkeler, D., Powers, A. C., Whitesell, R. R., Wang, T. G.: Water Soluble Polymers for Immunoisolation II: Evaluation of Multicomponent Microencapsulation Systems. Vol. 136, pp. 53-74.
Prokop, A., Hunkeler, D., DiMari, S., Haralson, M. A., Wang, T. G.: Water Soluble Polymers for Immunoisolation I: Complex Coacervation and Cytotoxicity. Vol. 136, pp. 1-52.
Pukánszky, B. and *Fekete, E.*: Adhesion and Surface Modification. Vol. 139, pp. 109-154.
Putnam, D. and *Kopecek, J.*: Polymer Conjugates with Anticancer Acitivity. Vol. 122, pp. 55- 124.

Quirk, R.P. and *Yoo, T., Lee, Y., M., Kim, J.* and *Lee, B.*: Applications of 1,1-Diphenylethylene Chemistry in Anionic Synthesis of Polymers with Controlled Structures. Vol. 153, pp. 67-162

Ramaraj, R. and *Kaneko, M.*: Metal Complex in Polymer Membrane as a Model for Photosynthetic Oxygen Evolving Center. Vol. 123, pp. 215-242.
Rangarajan, B. see Scranton, A. B.: Vol. 122, pp. 1-54.
Raphaël, E. see Léger, L.: Vol. 138, pp. 185-226.
Reddinger, J. L. and *Reynolds, J. R.*: Molecular Engineering of π-Conjugated Polymers. Vol. 145, pp. 57-122.
Reichert, K. H. see Hunkeler, D.: Vol. 112, pp. 115-134.
Rehahn, M., Mattice, W. L., Suter, U. W.: Rotational Isomeric State Models in Macromolecular Systems. Vol. 131/132, pp. 1-475.
Reynolds, J.R. see Reddinger, J. L.: Vol. 145, pp. 57-122.
Richter, D. see Ewen, B.: Vol. 134, pp.1-130.
Risse, W. see Grubbs, R.: Vol. 102, pp. 47-72.
Rivas, B. L. and *Geckeler, K. E.*: Synthesis and Metal Complexation of Poly(ethyleneimine) and Derivatives. Vol. 102, pp. 171-188.
Robin, J. J. see Boutevin, B.: Vol. 102, pp. 105-132.
Roe, R.-J.: MD Simulation Study of Glass Transition and Short Time Dynamics in Polymer Liquids. Vol. 116, pp. 111-114.
Roovers, J., Comanita, B.: Dendrimers and Dendrimer-Polymer Hybrids. Vol. 142, pp 179-228.

Rothon, R. N.: Mineral Fillers in Thermoplastics: Filler Manufacture and Characterisation. Vol. 139, pp. 67-108.
Rozenberg, B. A. see Williams, R. J. J.: Vol. 128, pp. 95-156.
Ruckenstein, E.: Concentrated Emulsion Polymerization. Vol. 127, pp. 1-58.
Rusanov, A. L.: Novel Bis (Naphtalic Anhydrides) and Their Polyheteroarylenes with Improved Processability. Vol. 111, pp. 115-176.
Russel, T. P. see Hedrick, J. L.: Vol. 141, pp. 1-44.
Rychlý, J. see Lazár, M.: Vol. 102, pp. 189-222.
Ryzhov, V. A. see Bershtein, V. A.: Vol. 114, pp. 43-122.

Sabsai, O. Y. see Barshtein, G. R.: Vol. 101, pp. 1-28.
Saburov, V. V. see Zubov, V. P.: Vol. 104, pp. 135-176.
Saito, S., Konno, M. and *Inomata, H.*: Volume Phase Transition of N-Alkylacrylamide Gels. Vol. 109, pp. 207-232.
Samsonov, G. V. and *Kuznetsova, N. P.*: Crosslinked Polyelectrolytes in Biology. Vol. 104, pp. 1-50.
Santa Cruz, C. see Baltá-Calleja, F. J.: Vol. 108, pp. 1-48.
Santos, S. see Baschnagel, J.: Vol. 152, p. 41-156.
Sato, T. and *Teramoto, A.*: Concentrated Solutions of Liquid-Christalline Polymers. Vol. 126, pp. 85-162.
Schäfer R. see Köhler, W.: Vol. 151, pp. 1-59.
Scherf, U. and *Müllen, K.*: The Synthesis of Ladder Polymers. Vol. 123, pp. 1-40.
Schmidt, M. see Förster, S.: Vol. 120, pp. 51-134.
Schopf, G. and *Koßmehl, G.*: Polythiophenes - Electrically Conductive Polymers. Vol. 129, pp. 1-145.
Schweizer, K. S.: Prism Theory of the Structure, Thermodynamics, and Phase Transitions of Polymer Liquids and Alloys. Vol. 116, pp. 319-378.
Scranton, A. B., Rangarajan, B. and *Klier, J.*: Biomedical Applications of Polyelectrolytes. Vol. 122, pp. 1-54.
Sefton, M. V. and *Stevenson, W. T. K.*: Microencapsulation of Live Animal Cells Using Polycrylates. Vol. 107, pp. 143-198.
Shamanin, V. V.: Bases of the Axiomatic Theory of Addition Polymerization. Vol. 112, pp. 135-180.
Sheiko, S. S.: Imaging of Polymers Using Scanning Force Microscopy: From Superstructures to Individual Molecules. Vol. 151, pp. 61-174.
Sherrington, D. C. see Cameron, N. R., Vol. 126, pp. 163-214.
Sherrington, D. C. see Lin, J.: Vol. 111, pp. 177-220.
Sherrington, D. C. see Steinke, J.: Vol. 123, pp. 81-126.
Shibayama, M. see Tanaka, T.: Vol. 109, pp. 1-62.
Shiga, T.: Deformation and Viscoelastic Behavior of Polymer Gels in Electric Fields. Vol. 134, pp. 131-164.
Shoda, S. see Kobayashi, S.: Vol. 121, pp. 1-30.
Siegel, R. A.: Hydrophobic Weak Polyelectrolyte Gels: Studies of Swelling Equilibria and Kinetics. Vol. 109, pp. 233-268.
Silvestre, F. see Calmon-Decriaud, A.: Vol. 207, pp. 207-226.
Sillion, B. see Mison, P.: Vol. 140, pp. 137-180.
Singh, R. P. see Sivaram, S.: Vol. 101, pp. 169-216.
Sivaram, S. and *Singh, R. P.*: Degradation and Stabilization of Ethylene-Propylene Copolymers and Their Blends: A Critical Review. Vol. 101, pp. 169-216.
Starodybtzev, S. see Khokhlov, A.: Vol. 109, pp. 121-172.
Steinke, J., Sherrington, D. C. and *Dunkin, I. R.*: Imprinting of Synthetic Polymers Using Molecular Templates. Vol. 123, pp. 81-126.
Stenzenberger, H. D.: Addition Polyimides. Vol. 117, pp. 165-220.
Stevenson, W. T. K. see Sefton, M. V.: Vol. 107, pp. 143-198.
Sumpter, B. G., Noid, D. W., Liang, G. L. and *Wunderlich, B.*: Atomistic Dynamics of Macromolecular Crystals. Vol. 116, pp. 27-72.
Sugimoto, H. and *Inoue, S.*: Polymerization by Metalloporphyrin and Related Complexes. Vol. 146, pp. 39-120

Suter, U. W. see Gusev, A. A.: Vol. 116, pp. 207-248.
Suter, U. W. see Leontidis, E.: Vol. 116, pp. 283-318.
Suter, U. W. see Rehahn, M.: Vol. 131/132, pp. 1-475.
Suter, U. W. see Baschnagel, J.: Vol. 152, p. 41-156.
Suzuki, A.: Phase Transition in Gels of Sub-Millimeter Size Induced by Interaction with Stimuli. Vol. 110, pp. 199-240.
Suzuki, A. and *Hirasa, O.*: An Approach to Artifical Muscle by Polymer Gels due to Micro-Phase Separation. Vol. 110, pp. 241-262.

Tagawa, S.: Radiation Effects on Ion Beams on Polymers. Vol. 105, pp. 99-116.
Tan, K. L. see Kang, E. T.: Vol. 106, pp. 135-190.
Tanaka, T. see Penelle, J.: Vol. 102, pp. 73-104.
Tanaka, H. and *Shibayama, M.*: Phase Transition and Related Phenomena of Polymer Gels. Vol. 109, pp. 1-62.
Tauer, K. see Guyot, A.: Vol. 111, pp. 43-66.
Teramoto, A. see Sato, T.: Vol. 126, pp. 85-162.
Terent´eva, J. P. and *Fridman, M. L.*: Compositions Based on Aminoresins. Vol. 101, pp. 29-64.
Theodorou, D. N. see *Dodd, L. R.*: Vol. 116, pp. 249-282.
Thomson, R. C., Wake, M. C., Yaszemski, M. J. and *Mikos, A. G.*: Biodegradable Polymer Scaffolds to Regenerate Organs. Vol. 122, pp. 245-274.
Tokita, M.: Friction Between Polymer Networks of Gels and Solvent. Vol. 110, pp. 27-48.
Tries, V. see Baschnagel, J:. Vol. 152, p. 41-156.
Tsuruta, T.: Contemporary Topics in Polymeric Materials for Biomedical Applications. Vol. 126, pp. 1-52.

Uyama, H. see Kobayashi, S.: Vol. 121, pp. 1-30.
Uyama, Y: Surface Modification of Polymers by Grafting. Vol. 137, pp. 1-40.

Vasilevskaya, V. see Khokhlov, A.: Vol. 109, pp. 121-172.
Vaskova, V. see Hunkeler, D.: Vol.:112, pp. 115-134.
Verdugo, P.: Polymer Gel Phase Transition in Condensation-Decondensation of Secretory Products. Vol. 110, pp. 145-156.
Vettegren, V. I.: see Bronnikov, S. V.: Vol. 125, pp. 103-146.
Viovy, J.-L. and *Lesec, J.*: Separation of Macromolecules in Gels: Permeation Chromatography and Electrophoresis. Vol. 114, pp. 1-42.
Vlahos, C. see Hadjichristidis, N.: Vol. 142, pp. 71-128.
Volksen, W.: Condensation Polyimides: Synthesis, Solution Behavior, and Imidization Characteristics. Vol. 117, pp. 111-164.
Volksen, W. see Hedrick, J. L.: Vol. 141, pp. 1-44.
Volksen, W. see Hedrick, J. L.: Vol. 147, pp. 61-112.

Wake, M. C. see Thomson, R. C.: Vol. 122, pp. 245-274.
Wandrey C., Hernández-Barajas, J. and *Hunkeler, D.*: Diallyldimethylammonium Chloride and its Polymers. Vol. 145, pp. 123-182.
Wang, K. L. see Cussler, E. L.: Vol. 110, pp. 67-80.
Wang, S.-Q.: Molecular Transitions and Dynamics at Polymer/Wall Interfaces: Origins of Flow Instabilities and Wall Slip. Vol. 138, pp. 227-276.
Wang, T. G. see Prokop, A.: Vol. 136, pp.1-52; 53-74.
Whitesell, R. R. see Prokop, A.: Vol. 136, pp. 53-74.
Williams, R. J. J., Rozenberg, B. A., Pascault, J.-P.: Reaction Induced Phase Separation in Modified Thermosetting Polymers. Vol. 128, pp. 95-156.
Winter, H. H., Mours, M.: Rheology of Polymers Near Liquid-Solid Transitions. Vol. 134, pp. 165-234.
Wu, C.: Laser Light Scattering Characterization of Special Intractable Macromolecules in Solution. Vol 137, pp. 103-134.
Wunderlich, B. see Sumpter, B. G.: Vol. 116, pp. 27-72.

Xiang, M. see Jiang, M.: Vol. 146, pp. 121-194.
Xie, T. Y. see Hunkeler, D.: Vol. 112, pp. 115-134.
Xu, Z., Hadjichristidis, N., Fetters, L. J. and *Mays, J. W.*: Structure/Chain-Flexibility Relationships of Polymers. Vol. 120, pp. 1-50.

Yagci, Y. and *Endo, T.*: N-Benzyl and N-Alkoxy Pyridium Salts as Thermal and Photochemical Initiators for Cationic Polymerization. Vol. 127, pp. 59-86.
Yannas, I. V.: Tissue Regeneration Templates Based on Collagen-Glycosaminoglycan Copolymers. Vol. 122, pp. 219-244.
Yamaoka, H.: Polymer Materials for Fusion Reactors. Vol. 105, pp. 117-144.
Yasuda, H. and *Ihara, E.*: Rare Earth Metal-Initiated Living Polymerizations of Polar and Nonpolar Monomers. Vol. 133, pp. 53-102.
Yaszemski, M. J. see Thomson, R. C.: Vol. 122, pp. 245-274.
Yoo, T. see Quirk, R.P.: Vol. 153, pp. 67-162
Yoon, D. Y. see Hedrick, J. L.: Vol. 141, pp. 1-44.
Yoshida, H. and *Ichikawa, T.*: Electron Spin Studies of Free Radicals in Irradiated Polymers. Vol. 105, pp. 3-36.

Zhou, H. see Jiang, M.: Vol. 146, pp. 121-194.
Zubov, V. P., Ivanov, A. E. and *Saburov, V. V.*: Polymer-Coated Adsorbents for the Separation of Biopolymers and Particles. Vol. 104, pp. 135-176.

Subject Index

A₂B₂ heteroarm, star polymers 146
ABC-type block copolymers 129
Absorption function of sensor 196
Acetaldehyde 38
Adhesives 47
Alkoxide-modified initiator 138
Alkyl methacrylates 76
Alkyllithium aggregates 72
Allyl carbanion 76
Alternating copolymer 7
Amide functionalization 114
Anisole 133
Annealing 54
Anthracene labeling 120, 121

Band gap tailoring 174
Base pairing 13
Base stacking 10
Benzyl carbanion 88
Benzyllithium 74
Bimodal distribution 136
Bioadhesives 53
Biocompatibility 58
Biomaterials 58
1,1-Bis(4-dimethylaminophenyl)ethylene 110
Bis(dimethylamino)polystyrene 109
Bis(trimethylsilyl)amine group 113
Bovine serum albumin 59
BSA 59
Bulk-heterojunctions 179
Butadiene copolymerization 99, 122
tert-Butoxide 138
tert-Butyl acrylate 80
tert-Butyl methacrylate 125
tert-Butyllithium 72
sec-Butyllithium initiator fragment 89

Carbon acids 75, 76
Carbon indicator 75
Carbonation 102
Charge transfer complex 7
- - -, reaction 7
Circular dichroism (CD) 9

Co-condensation synthesis method 173
Conductivity, percolation behavior 175
Connectivity, percolating networks 177
Contact lens 60
Coplanarity 77, 79, 117
Copolymerization 6
Core-shell nanoparticles 170
Coupling reaction 3
Crosslinking agent 39
Crosslinking density 41
Crosslinking, physical 40
Cryochemical synthesis 173
Crystallinity 50
-, degree 43
-, drying 42
Crystallites 41, 48
Crystallizability 38
Crystallization 47
4-Cyanostyrene 117
p-Cyclophane 172

Debye length 178
Depurination 22
Depyrimidination 22
Diaddition 72
Diaminopolystyrene 110
1,1-Diarylethylene 69
Dibromoxylene 112
Dielectric constant anomaly 198
Diene copolymerization 99, 122
Diene initiation 80
Diene microstructure 80, 101, 136, 137
Differential scanning calorimetry 48
Diffused p-n junctions 177
Diffusion 59
Diisopropylamide group 114
Dilithium initiators 82, 83, 132
1,4-Dilithium-1,1,4,4-tetraphenylbutane 82
Diphenolpolystyrene 116
1,1-Diphenylalkyllithium 74
-, adduct 85
-, aggregation 74
-, hybridization 77

1,1-Diphenylethylenes, ring-substituted 86
Diphenylmethyl carbanion 81
Dispersed anode (hole collector) 194
Dissolution-controlled system 54
Disulfonatopolystyrene 104
DNA templating nanocomposite synthesis 203
Donor-acceptor electron transfer 179
DPE addition, rate 87
-, PLi addition kinetics 84
-, reactivity 74
Drug delivery 51

End-capping 91, 102, 130
Energy conversion efficiency, solar cell 186
Epichlorohydrin 104
Equivalent network resistance 195
Ethanol sensor 198
Ethylene glycol 43
Excimer 12
Exciton and charge transfer 192

Fractalization of space charge layer 167, 177
Franc-Keldysh effect 175
Freezing/thawing 46, 55

Gas sensor Field Effect Transistors (GasFET) 198
Gels, freeze/thawed 46
Giant cross capacitance 198
Gradient composites 169
Growth factors, controlled release 53

Hammett values 87, 108, 113, 116, 117, 122
Hammond postulate 88, 89
Hexamethyltrisiloxane 126
Hopping conduction 176
- -, Schklovsky-Efros model 196
Humidity nanocomposite sensors 197
Hybrid orbital 78
Hybridization 17
Hydrodynamic volume 151
Hydrogen bonding, intermolecular 38
Hydrogen sensor 198
Hydrolysis 38
Hyperchroism 10
Hypochroism 10

In situ synthesis reactions 167
Information storage density problem 201
Initiators 79, 93, 132, 138
Integral brightness of LED 180
IR spectroscopy 44
γ-Irradiation 39

Irradiation dose 40
Isoprenyldilithium 83

LED, external quantum efficiency 180
-, peak brightness 181
Lewis base 75, 92, 93, 137, 140
^{7}Li NMR 92
Lithium alkoxide 92
Lithium sec-butoxide 139
Lithium butyldimethylsilanolate 80
Lithium dihydronaphthylide 132
Lithium methoxide 80
Lithium 2-(2-methoxy)ethoxide 80
Lithium-halogen exchange 130

Magnetic anisotrpoy energy 199
Magnetic force microscopy 200
Magnetic random access memory 199
Magnetization measurements 200
Mayo-Lewis copolymerization equation 96
MDDPE 127, 129, 130, 133, 143
Melting temperature (T_m) 15
Metal-insulator transition 175
Microparticles 51
Mikto-arm 142
Modification of nanopaticle surface 171
Mössbauer spectroscopy, nanocomposites 200
Mott transition 175
Multi-cycle treatment synthesis 168

Nanocomposites, superparamagnetic behavior 199
-, Mössbauer spectroscopy 200
Nanoparticles, capping 172
-, ordered arrays, polymer matrix 203
Naphthalene labeling 118-121
Network, crosslinked 39
Neural networks, polymer-nanocomposites 203
Number average molecular weight 38

Onsager dissociation 186
Open circuit voltage, solar cell 186
Optical rotary dispersion (ORD) 9
Ordered arrays, nanoparticles, polymer matrix 203
Organolithium aggregation 72
Organolithium initiator, trifunctional 132
Organometallic monomers 172
Oxazolyl-functionalized polymers 114

p-i-n device structure 188
p-n nanojunctions 167
PAAm 61

Subject Index

PDDPE 126, 134
PEG 61
Penultimate effect 97
PEO 57
Peptide nucleic acid (PepNA) 17
Percolation behavior, conductivity 175
Phenanthrene labeling 121
Phenol functionalization 108
Photoconductivity 184
Photodiode, spectral current sensitivity 185
Photoinduced charge transfer 186
Photoluminescence quenching 187
Photoresist 31
Photovoltaic effect 185
Plastic solar cells 186
PNAs, catalytic activity 24
Poly(butadienyl)lithium 91, 92
Poly(2,3-dimethylbutadienyl)lithium 100
Poly(dienyl)lithium 84
-, aggregation 90, 92, 93, 100
Poly(ethylene glycol) 61
Poly(ethylene oxide) 57
- macromonomer 129
Poly(2,4-hexadienyl)lithium 90, 93
Poly(isoprenyl)lithium 91, 92
Poly(o-methoxystyrene), isotactic 98
Poly(styrene-*block*-butadiene) 111
Poly(styryl)cesium 87
Poly(styryl)lithium 84, 89
Poly(vinyl acetate) 38
Polyacrylamide 61
Polybutadiene macromonomer 128
Polydimethylsiloxane 125
Polyelectrolytes 9
Polymerisation in glow discharge 173
Polynucleotide analogues (PNA) 1, 2
Potassium alkoxide 120
1,3-Propanesulfone 107
PVA, freezing/thawing 45, 47, 52
- gels 43, 48, 49
- grades 38
- hydrogels 45, 50
- -, crystalline 40
- membranes, semicrystalline 52
- microparticles 51
-, semicrystalline 54
Pyrene-labeling 119, 121

Quantitative end-capping 89
Quantum confinement 173

Rate constants 96
γ–Rays 39
Refractive index tailoring 167

SBS, linear 151
Schklovsky-Efros model, hopping conduction 196
Self-assembly 153
- synthesis 170
Sensor arrays 196
Sensor transducre function 196
Short circuit current, solar cell 186
Silicon tetrachloride 148
Single electron transistor 203
Solar cell, energy conversion efficiency 186
- - fill factor 186
Solubility 38
Solute transport 52
Space charge layer formation 190
Spin electronics 199
Spin-casting method, nanocomposite film preparation 169
Spintronics 199
Stability, long-term 166
Stratification 170
Styrene initiation 80
Sub-light sources 204
Sulfides, chemical conversion 168
Superparamagnetic behavior, nano-composites 199
Supramolecular engineering 153
Swelling behavior 50
- degree 41
Swelling-induced conductivity change 196

Template polymerization 30
Tensile properties 97, 150
Tetrabutylammonium fluoride 131
Tetrahydrofuran 92
Tetramer-dimer equilibria 74
1,1,3,3-Tetraphenyloctane 71
Triethylamine 123, 133
Trimethylsilyl group labeling 117
Twisted ring 77

Viscosity 39
Voltage-depended emission color 183

Winstein spectrum 89

X-ray analysis 43

Printing: Saladruck, Berlin
Binding: Lüderitz & Bauer, Berlin

RETURN TO: CHEMISTRY LIBRARY
100 Hildebrand Hall • 642-3753

LOAN PERIOD 1	2	3
4	2 HOUR	

DUE AS STAMPED BELOW.

NON-CIRCULATING UNTIL: 11/17/00		
DEC 20 2001		
MAY 24 2003		
JAN 05 '04		

FORM NO. DD 10
3M 3-00

UNIVERSITY OF CALIFORNIA, BERKELEY
Berkeley, California 94720–6000